Management of Hydrological Systems

Management of Hydrological Systems

Analysis and Perspective of the Contingent Valuation of Water for Mountain Basins

Holger Manuel Benavides Muñoz
Universidad Técnica Particular de Loja, Loja, Ecuador

Jorge Eugenio Arias Zari

Andreas Erwin Fries
Universidad Técnica Particular de Loja, Loja, Ecuador

José Sánchez-Paladines
Universidad Técnica Particular de Loja, Loja, Ecuador

Antonio Jesús Gallegos Reina
Universidad de Málaga, Malaga, Spain

Raquel Verónica Hernández Ocampo

Pablo Ochoa Cueva
Universidad Técnica Particular de Loja, Loja, Ecuador

CRC Press
Taylor & Francis Group
Boca Raton London New York Leiden

CRC Press is an imprint of the
Taylor & Francis Group, an **informa** business

A BALKEMA BOOK

CRC Press/Balkema is an imprint of the Taylor & Francis Group, an informa business

© 2020 Taylor & Francis Group, London, UK

Typeset by Apex CoVantage, LLC

Library of Congress Cataloging-in-Publication data
Applied for

Published by: CRC Press/Balkema
Schipholweg 107C, 2316 XC Leiden, The Netherlands
e-mail: Pub.NL@taylorandfrancis.com
www.crcpress.com – www.taylorandfrancis.com

ISBN: 978-0-367-45655-9 (Hbk)
ISBN: 978-1-003-02457-6 (eBook)
DOI: 10.1201/9781003024576
https://doi.org/10.1201/9781003024576

Contents

List of tables		vii
List of figures		ix
Affiliations of authors		xiii
Acknowledgments		xv

1 Introduction   1

HOLGER BENAVIDES-MUÑOZ, JORGE ARIAS-ZARI AND JOSÉ SÁNCHEZ-PALADINES

1.1	Justification and problems	1
1.2	Objectives	3
1.3	Hypotheses	3

2 Baseline of study   5

ANDREAS FRIES, HOLGER BENAVIDES-MUÑOZ, ANTONIO GALLEGOS REINA, JORGE ARIAS-ZARI, JOSÉ SÁNCHEZ-PALADINES AND PABLO OCHOA-CUEVA

2.1	Hydrological system	5
2.2	Climate of Ecuador	9
2.3	Climate of Loja	15
2.4	Solar radiation	18
2.5	Air temperature	20
2.6	Potential evapotranspiration	22
2.7	Relative humidity	22
2.8	Precipitation	24
2.9	Hydrologic balance	26
2.10	Case of hydrological study	27
2.11	Hydrological-hydraulic study and risk of flooding in the Zamora Huayco basin	39
2.12	Soil erosion	55

3   **Market study on drinking water**   61
    HOLGER BENAVIDES-MUÑOZ, JORGE ARIAS-ZARI AND JOSÉ SÁNCHEZ-PALADINES

    3.1   Universe and samples in market studies   61
    3.2   Characteristics of the surveys applied in market studies   63

4   **Contingent valuation of the water service**   69
    HOLGER BENAVIDES-MUÑOZ, JORGE ARIAS-ZARI AND JOSÉ SÁNCHEZ-PALADINES

    4.1   Independent variables (categorical and continuous) vs.
          the dependent variable   69
    4.2   Financial evaluation – environmental   76

5   **Background and environmental problems**   87
    RAQUEL HERNÁNDEZ-OCAMPO

    5.1   Environmental problems in the southern region of Ecuador   89
    5.2   Demand for water in El Carmen and San Simón micro-basins   93
    5.3   Perspectives on this natural environment   95

6   **Climate change**   97
    ANDREAS FRIES

    6.1   General overview   97
    6.2   Climate change in Loja   101

7   **Conclusions and recommendations**   113
    HOLGER BENAVIDES-MUÑOZ

    7.1   Conclusions   113
    7.2   Recommendations   114

    Bibliography   115
    Annexes   121

Tables

2.1	Meteorological stations and available station data	17
2.2	Perimeter and surface of the El Carmen and San Simón micro-basins	27
2.3	Tendencies in floods according to Kc	29
2.4	Gravelius coefficient (Kc)	29
2.5	Calculation of the mean slope of the El Carmen basin	30
2.6	Calculation of the average slope of the San Simón basin	30
2.7	Average percentage slope (Sc)	30
2.8	Global slope index of P. Dubré, in m/km (Ig)	31
2.9	Type of relief based on Ig	32
2.10	Specific gradient (Ds)	32
2.11	Type of relief, depending on the Ds	32
2.12	Density of the network (Dr) in riverbeds/km^2	33
2.13	Density of drainage, in km/km^2	33
2.14	Average elevation of El Carmen micro-basin (graphic method)	35
2.15	Average elevation of El Carmen micro-basin (numerical method)	36
2.16	Average elevation of the San Simón micro-basin (graphic method)	38
2.17	Average elevation of the San Simón basin (numerical method)	39
2.18	Generation of precipitations (isohyet lines)	39
2.19	Summary of generated precipitations	40
4.1	Ratio gen vs. aip (US$)	70
4.2	Ratio al vs. aip (US$)	70
4.3	Relationship age and aip (US$)	70
4.4	Ratio ardw vs. aip (US$)	70
4.5	Ratio arvcb vs. aip (US$)	71
4.6	Ratio arbqws vs. aip (US$)	71
4.7	Ratio arbqlws vs. aip (US$)	71
4.8	Ratio arbrws vs. aip (US$)	71
4.9	Relation arlpo vs. aip (US$)	72
4.10	Relation arbmow vs. aip (US$)	72
4.11	Relation arbr vs. aip (US$)	73
4.12	Relation occup vs. aip (US$)	73
4.13	Relation ninh vs. aip (US$)	73

4.14   Ratio mpws vs. aip (US$)   74
4.15   Relation nha vs. aip (US$)   74
4.16   Relation mfi vs. aip (US$)   74
4.17   Relation mfe vs. aip (US$)   75
4.18   Distribution of variables according to age   75
4.19   Analysis of contingency between age and gender   77
4.20   Contingency analysis between age and academic level   78
4.21   Contingency analysis between age and TAPPA   79
4.22   Contingency analysis between age and ARBMWRP   80
4.23   Analysis of contingency between age and ARBR   81
4.24   Price example for the water service   82
4.25   Characteristics of endowments and fees for water service   82
6.1   Average monthly T and P values for the period 1966–1995   111
6.2   Average monthly T and P values for the period 1986–2015   111

Figures

2.1   Diagram of the drainage basin of the Amazon river   6
2.2   Outline of the Amazonas and Hamza rivers   6
2.3   Map of the Santiago river water contributions   7
2.4   Image of the Zamora river in Zumbi city – Ecuador   8
2.5   Image of the Zamora river in Zamora city – Ecuador   8
2.6   Image of the Zamora river – Ecuador, seen from the bridge of the Zamora city   8
2.7   Image of the Zamora river in Zamora city – Ecuador   8
2.8   Image of the Zamora Huayco and Malacatos rivers union in the city of Loja – Ecuador   9
2.9   Abbreviated topological diagram of the Zamora river   10
2.10   Image of the junction of the San Simón and El Carmen streams, Loja – Ecuador   11
2.11   Climatic regions of Ecuador   11
2.12   (a) Global atmospheric circulation, (b) global ocean currents   12
2.13   Climate graph for Guayaquil (coast)   13
2.14   Time of vertical sun position at each latitude   14
2.15   Climate graph for Quito (Sierra)   15
2.16   Climate graph for Puyo (Amazon basin)   16
2.17   Digital elevation model (DEM) of the city of Loja, including climate stations   17
2.18   Climate graph for the city of Loja (station Argelia 1965–2015)   18
2.19   Average solar radiation for Loja: (a) July, (b) November and (c) year, based on information from 2011 to 2017   19
2.20   Monthly average minimum (a), mean (b) and maximum (c) temperatures for November, based on information from 2011 to 2017   21
2.21   Monthly average minimum (a), mean (b) and maximum (c) temperatures for July, based on information from 2011 to 2017   21
2.22   Average potential evapotranspiration for Loja: (a) March, (b) September and (c) year, based on information from 2011 to 2017   22
2.23   Monthly average minimum (a), mean (b) and maximum (c) RH for February, based on information from 2011 to 2017   23

2.24 Monthly average minimum (a), mean (b) and maximum (c) RH for
      September, based on information from 2011 to 2017   24
2.25 Average precipitation amounts in Loja: (a) March, (b) September
      and (c) year, based on information from 2011 to 2017   25
2.26 Average water surplus or water deficit for Loja: (a) March,
      (b) September and (c) year, based on information from 2011 to 2017 26
2.27 Diagram of the El Carmen micro-basin   28
2.28 Diagram of the San Simón micro-basin   28
2.29 Longitudinal profile of the Quebrada El Carmen   34
2.30 Longitudinal profile of the Quebrada San Simón   34
2.31 Hypsometric curve for El Carmen micro-basin   36
2.32 Hypsometric curve for San Simón micro-basin   37
2.33 Precipitation–altitude relationship   40
2.34 Three-dimensional representation of the drainage basin corresponding
      to the section examined   42
2.35 Location of the stations used for rainfall analysis and UTM
      and altitude data   43
2.36 Flow calculation table according to the modified rational method   45
2.37 Flooding sheet resulting in the neighborhood of the Zamora
      Huayco district and previous section   47
2.38 Depth of flow   48
2.39 Flow speed   49
2.40 Cutting flow stress   50
2.41 Water surface elevations and cross sections of the control
      sections of the sector of the studied channel   51
2.42 Accumulation of heterometric sediments and grounding
      in works of the passage of a stream   55
2.43 Digital elevation model (DEM) of the city of Loja, including
      the Zamora Huayco watershed   56
2.44 Land use in 2008 in the Zamora Huayco watershed   57
2.45 Potential soil erosion [t/ha] in the Zamora Huayco watershed,
      based on vegetation data from 2008   58
2.46 (a) Projected vegetation map for 2040 (right) and (b) the resulting
      potential soil erosion [t/ha] (left) in the Zamora Huayco watershed 59
3.1   Hydraulic delimitation of La Tebaida hydrometric district   62
3.2   Number of people per dwelling   63
3.3   Inhabited premises   63
3.4   Monthly representation by water bills   63
3.5   Monthly payment for water consumption   63
3.6   Time in hours of the available water service per day   64
3.7   Hours of water service per day   64
3.8   Characteristics of the counter   64
3.9   State of the counter   64
3.10 Problems perceived in the service   65
3.11 Perception of service problems   65
3.12 Solution to perceived problems   65
3.13 Proposed solution to problems   65

3.14 Incremental payment availability 66
3.15 Willingness to pay the additional value 66
3.16 Curve of positive externality in consumption 67
6.1 Observed surface temperature change from 1901 to 2012, derived from
temperature trends determined by linear regression 98
6.2 Observed precipitation change from 1901 to 2010 and from 1951 to 2010 98
6.3 Extent of Arctic summer (July–September) average sea ice 99
6.4 Global mean sea level relative to the 1900–1905 mean 99
6.5 Distribution of net emissions in Ecuador in 2012 100
6.6 Distribution of net emissions in Ecuador in 2012 101
6.7 Yearly temperature trends over the past 50 years for the city of Loja 102
6.8 Deviated trends of daily Tmin data: (a) TNN, (b) TNX, (c) TN10p
and (d) TN90p 102
6.9 Deviated trends of daily Tmax data: (a) TNN, (b) TNX, (c) TN10p
and (d) TN90p 105
6.10 Trends for annual precipitation: (a) personal calculation and
(b) RClimDex 1.0 108
6.11 Deviated trends of daily P data: (a) R10mm, (b) R20mm
and (c) R95p 109

Affiliations of authors

Holger Benavides-Muñoz, Professor of Water Resources, Universidad Técnica Particular de Loja. San Cayetano Alto, 110107, Loja, Ecuador. E-mail: hmbenavides@utpl.edu.ec

Jorge Arias-Zari, Freelance Economist. Economy and Projects Management, Loja, Ecuador. E-mail: gerardo01958@hotmail.com

Andreas Fries, Professor of Water Resources, Universidad Técnica Particular de Loja. San Cayetano Alto, 110107, Loja, Ecuador. E-mail: aefries@utpl.edu.ec

José Sánchez-Paladines, Visiting Professor of Universidad Técnica Particular de Loja. San Cayetano Alto, 110107, Loja, Ecuador. E-mail: jasanchez@utpl.edu.ec

Antonio Gallegos Reina, Visiting Professor of Universidad Técnica Particular de Loja. Universidad de Málaga, Campus de Teatinos, C.P. 29071, Málaga, Spain. E-mail: a.gallegos@uma.es

Raquel Hernández-Ocampo, Master in Environmental Administration. Loja, Ecuador. E-mail: raqverho4@yahoo.es

Pablo Ochoa-Cueva, Professor of Natural Resources, Universidad Técnica Particular de Loja. San Cayetano Alto, 110107, Loja, Ecuador. E-mail: paochoa@utpl.edu.ec

Acknowledgments

The authors acknowledge their gratitude to the Universidad Técnica Particular de Loja for the support provided for the development of this document; to Eng. Galo Paltín Saraguro for his participation in the study of cadaster and redesign of the La Tebaida subsector; to MSc. Franz Pucha Cofrep for his contribution to the generation of Figure 2.3 and to Dr. Fernando Oñate-Valdivieso for his contribution to the generation of two images of the study sector included in the annexes, through GIS tools.

Also, we thank all our friends and colleagues, who in one way or another, at a specific time, contributed to this work; and anonymous reviewers and Editors for their constructive comments.

Chapter 1

Introduction

*Holger Benavides-Muñoz, Jorge Arias-Zari
and José Sánchez-Paladines*

The present work consists of seven chapters: the first one refers to the opening and an explanation of the study context; the second exposes the physical-hydric characterization of the San Simón and El Carmen micro-basins; the third compiles the socioeconomic study of consumers in the studied areas of Loja, which consists of two phases, one carried out in the period 2004–2005 and the other in 2011, whose results are not coincident (the reasons will be explained later); the fourth contains the variables and contingent valuation of the water service; the fifth chapter contains environmental problems in the southern region of Ecuador, demand for water and perspectives on the natural environment; the sixth chapter is about climate change, and the seventh chapter contains the conclusions and recommendations.

It became necessary to use strategies and technical tools and apply them to an analysis of the sustainability of the water resource and the assessment of the water supply service in the city, as well as the analysis of stakeholders, market study, analysis of problems and objectives, financial analysis and diagnosis of contingency by using categorical variables. These were synthesized in an indicator-based management system called the Supply Sustainability Index – ISS (Benavides-Muñoz, 2018; Benavides-Muñoz, 2010).

The present work constitutes a technical contribution for decision-making involved in improving users' well-being served by the system of supply that the local company offers and the sustainable management of affluents and micro-basins.

The procedures proposed in this book can be applied to river basins whose characteristics are similar to those of the present study.

1.1 Justification and problems

Water basins give clues about their imminent water instability, which is reflected in a growing torrential regime and increasingly prolonged periods of drought.

The increase in the population in Latin America, and in particular in the city of Loja, Ecuador, has caused severe changes in the environment and has accelerated the degradation of natural resources in the river basins that supply water to the population, which are part of the Amazonian basin region. As a result, the demand for water resources is far exceeding the state's capacity to satisfy them, and furthermore, coverage rates in drinking water and sanitation services have become among the lowest in Latin America. Also, there are severe seasonal energy deficits that add to conflicts over the use of water.

The inadequate management of the hydrological basins causes adverse effects in biophysical aspects such as loss of soil, loss of biodiversity, alteration of the runoff volume in rivers and streams that generate substantial economic losses due to floods, restriction in agricultural production and diseases in humans, animals and crops, which progressively decreases the life quality of the region's inhabitants.

Due to such circumstances, the existence of extensive and valuable forests that expose extensive agricultural soils to erosion is threatened, and in recent times, large areas of valuable land for agriculture must have been abandoned, which causes a markedly high emigration of places that previously were considered productive. The evidence of this situation is manifested in high levels of water pollution, marked and increasingly long periods of drought, high rates of deforestation, accelerated erosion processes and invigoration in the regime of torrential runoff.

The hydrographic basin and particularly the water resources play crucial roles in the participation and integration of the actors involved in development and sustainability goals. Managing natural resources in river basins requires an integrated system approach, which includes a social factor, rural and urban areas and an agent that mobilizes actions and interventions on the environment and nature.

The future role of the effective and efficient use of water resources will be fundamental to the elimination of structural deficiencies in Ecuador's economy. The high rate of population migration from our province of Loja to the capital or eastern Ecuador provinces, a result of the drought in the 1970s, is an indicator that our region is one of the poorest in the country.

The interactions of the territorial system require a comprehensive and interdisciplinary vision. A basin that is not adequately managed affects the cities located in it or that depend on it. A lack of water and poor water quality are fundamentally subordinated to the management of the integral water system of the source basin.

There have always been problems of contamination by agrochemicals, dumping of waste, dragging of solids and anthropic actions in general, which cause adverse environmental effects that lead to a high cost of water treatment for human consumption. Likewise, the decrease in the energy productivity of hydroelectric power plants is because in the receiving basin, there have been alterations in the vegetation cover that cause a variation of the runoff volume and also an abundant sediment transport that reduces the storage capacity of the reservoirs.

As an example, we point to the Paute hydroelectric project, in which the excessive erosion of the basins, or landslides, have notably reduced the storage volume of the vessel, which has caused a considerable reduction in its useful life.

In this context, it is imperative to manage natural drains technically and methodologically, in accordance with properly managing water basins, to control and reverse the degradation of natural resources through decisions at different levels of urban–rural development projects.

This proposal is aimed at studying the sustainability of the water supply system for Loja, fed by the water of the El Carmen and San Simón streams and protection of its environment and biodiversity of the basins that provide water to these streams.

The issue has to do with the environmental economy and contingent valuation of water that mark the sustainability of the project and is also related to the analysis of impacts generated by the reforestation and maintenance of the affluent micro-basins of El Carmen and San Simón, concerning the quality and quantity of water.

1.1.1   *Analysis of the problem*

Many projects for the exploitation of natural resources have been launched under the assumption that ecologically benign exploitation can reconcile economic growth and improvements in quality of life.

The accelerated degradation of natural resources and the environment itself, deforestation in areas of poor soils, steep areas along rivers and increase in population, poverty and socio-environmental insecurity comes with a high risk of vulnerability to natural disasters, floods and soil erosion, which characterize most of the areas of our region. Deforestation, overgrazing, inadequate use of land, industrial and domestic agricultural practices, the scarcity of guidelines and ordinances of local governments, policies and integrated actions have led to a worrying state of deterioration and dismemberment in most of the basins in our country, with effects that are already evident in the reduction in the availability and quality of water, in soil erosion and in the increase in vulnerability to natural disasters.

This situation demands the immediate action of technicians able to participate in the management and control of basins to integrate the biophysical dimensions with the socio-economic ones in order to protect the environment. This requires modern approaches, strategies and knowledge on the participatory technical management of basins to achieve an adequate use and exploitation of natural resources and the recovery of those that have already been damaged.

In this study, the Zamora Huayco basin has a tremendous human impact on the exploitation of aggregates and agricultural use, which diminishes the natural spaces that preserve the equilibrium conditions of the environment. Additionally, exotic species that affect the ecosystem are introduced. The lack of integration between forest policy within the framework of a national environmental regime, progressive deforestation in the face of the uncontrolled advance of road infrastructure in protected areas and natural forests, the appropriation or invasion of forest lands by colonization processes and migration, insecurity in land tenure, non-forestry (agricultural) uses of immediate yield (not sustainable) and forest uses for logging purposes destructively and aggressively affect the forest. Also, the socio-economic contribution to the sector by the similar institutions is insufficient. There are shortcomings in current planning and there is a deficient or almost null incentive system, which adds misinformation and obsolete procedures in applying environmental protection laws.

An alternative to solve these problems arises in the present research.

1.2 Objectives

1.2.1 Overall objective

This project aims to determine the sustainability of the water resource utilization system from the El Carmen and San Simón micro-basins belonging to the Zamora Huayco sub-basin, by using economic-statistical techniques.

1.2.2 Specific objectives

- Project a payment rate for water consumption in Loja.
- Assign an economic value to the water resource through contingency analysis.
- Analyze the economic sustainability of the water production and treatment system, by using Net Present Value (NPV), Internal Rate of Return (IRR) and Benefit-Cost Ratio ($R^B/_C$).

1.3 Hypotheses

- The demanding social class (users of the potable water system) with an incremental availability of a monthly payment for clean water between US\$3 and US\$5 have a university education, are between 17 and 40 years old and consider that the fee for the water service is low and that the water quantity and quality are terrible and/or deficient.

- The demanding social class with maximum incremental payment availability (IPA) per payroll is one whose income exceeds US$ 1000 per month.
- The most significant social level of users will determine that the sustainable use of water from the micro-basins El Carmen and San Simón, belonging to the sub-basin of Zamora Huayco-Loja, is in relation to the municipal unit of intervention for drinking water and sewerage (UMAPAL, because of its initials in Spanish) but would in a more efficient and effective way be in its management of the control and the IPA of drinking water users.

1.3.1   Support tools

- Documents and methods that allow for establishing the baseline.
- An analysis of the cause and effects of problems, together with an analysis of means and ends of objectives.
- A market study through tabulated surveys.
- A comparison of the sample study variables (of the market) with the incremental availability of payroll payment of each petitioner.
- A contingency matrix.
- Financial evaluations (NPV, IRR and $R^B/_C$).

1.3.2   Sustainability strategy

This project becomes the bridge between human activity in micro-basins, environmental sanitation and water service to the people from Loja, in a sustainable manner. The rate of the permanent sustainability of the basin will be compensated for by the effectiveness of the management and the updated and sufficient information available.

Chapter 2

Baseline of study

Andreas Fries, Holger Benavides-Muñoz,
Antonio Gallegos Reina, Jorge Arias-Zari,
José Sánchez-Paladines and Pablo Ochoa-Cueva

The preparation of the baseline of this study is based on existing information, gathered from multiple sources, such as the "Study on erosion and protection of micro-basin of Zamora Huayco provider of drinking water for the city of Loja" by García and Ojeda (1994), authors from the Agricultural Research Institute of the Loja National University; another fundamental source is the work of Benavides and Solano (2005) titled "Evaluation of the environmental impact of afforestation and reforestation plans executed in the Zamora Huayco basin of the Loja canton, Loja province"; finally, the baseline relies on data provided by (*Unidad Municipal de Agua Potable y Alcantarillado de Loja*) UMAPAL and the Financial Department of the Municipality of Loja. As analytical-deductive academic methods, traditional procedures for the generation of hydrological and financial information will be used.

2.1 Hydrological system

2.1.1 Brief physical description of the Amazon basin

The Amazon river basin, which has an area of approximately 6.2 million km^2, is made up of several territories in several countries: Brazil (with an area of about 4 million km^2), Ecuador, Bolivia, Colombia, Guyana and Perú, and can be considered as the most significant hydrological basin in the world (see Fig. 2.1).

According to Foundation Bustamante De La Fuente (2010), in his document "Climate Change in Perú – Amazon region," the approximate drain of the Amazon river is 210,000 m^3 s^{-1}, which significantly increases during the rainy season.

From the doctoral thesis of Elizabeth Tavares Pimentel, directed by Prof. Valiya Hamza of the Amazonas Federal University, it can be inferred that there are indications that under the Amazon riverbed, there is an underground river named "Hamza" that flows at deficient speed, that has 6992 km of route: its width varies between 200 km and 400 km; its depth varies between 2 km and 4 km, a zone in which the flow changes from vertical to horizontal; and its speed fluctuates between 10 m and 100 m per year (see Fig. 2.2).

According to the same source, the Hamza river is filled with waters from the Ecuadorian, Colombian and Perúvian Andes and poured into the depths of the Atlantic, about 200 km from the coast.

Figure 2.1 Diagram of the drainage basin of the Amazon river
Source: https://bit.ly/2OzopJW

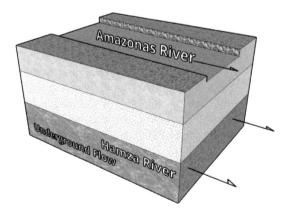

Figure 2.2 Outline of the Amazonas and Hamza rivers
Source: https://bit.ly/2QpmsSa

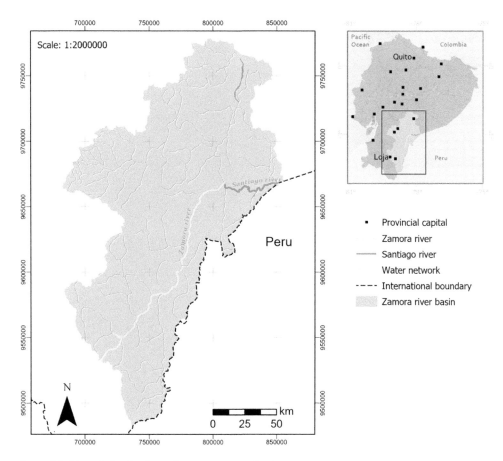

Figure 2.3 Map of the Santiago river water contributions

2.1.2 Main affluents of the Amazon river

Among the most considerable affluents of the Amazon is the Marañón river, whose principal affluents are as follows:

- Santiago (Fig. 2.3), Morona, Pastaza, Tigre, Napo, Putumayo, Caquetá, Japurá, Negro, Urubu, Uatama, Nhamundá, Trombetas, Curuá, Maicurú, Paru, Jarí and Maracá.
- Huallaga, Ucayali, Javari, Jandiatuba, Jutaí, Juruá, Tefé, Coari, Purus, Madeira, Paraná-Urariá, Tapajós, Xingú, Anapu, Pacaja and Jacunda.

Affluents of the Santiago river

The Santiago river is formed in Ecuador by the affluents of the Ecuadorian rivers Zamora and Namangoza and runs through the Amazonian territory of Ecuador and Perú.

Figure 2.4 Image of the Zamora river in Zumbi city – Ecuador

Figure 2.5 Image of the Zamora river in Zamora city – Ecuador

Figure 2.6 Image of the Zamora river – Ecuador, seen from the bridge of the Zamora city

Figure 2.7 Image of the Zamora river in Zamora city – Ecuador

Affluents of the Zamora river

Zamora river has as main affluents the Yacuambi and Nangaritza rivers (Figs. 2.4, 2.5, 2.6 and 2.7).

Zamora river in its origin

Zamora river has its origin in the Podocarpus National Park, specifically in the Cajanuma knot, at more than 3300 m of altitude in the limit of the Loja and Zamora Chinchipe Ecuadorian provinces, with the name Malacatos river, the same that has as affluents the Zamora Huayco river and the Jipiro river, among others, and is heading northeast, with the name Zamora river (see Fig. 2.8).

Malacatos river

Curitroje and Mónica streams are born in the Cajanuma knot and join in its spur at the point known as Dos Hermanos, giving rise to the "Malacatos river that goes to the Atlantic." This quotation requires the following explanation: in the same Cajanuma knot, in the southern

Figure 2.8 Image of the Zamora Huayco and Malacatos rivers union in the city of Loja – Ecuador

spur, "the Malacatos river that goes to the Pacific" is born (Ochoa *et al.*, 2017); the same moves downstream together with other rivers and streams form the critical Catamayo river, which crosses most of the province of Loja and later, before its mouth in the Pacific, takes the name Chira river, which is binational, between Ecuador and Perú.

This is how the Cajanuma knot constitutes the birth site and at the same time the *divortium aquarum* for the Malacatos river, which goes to the Pacific, and the Malacatos river, which goes to the Atlantic. The Malacatos river, which initially has its path to the north in the city of Loja, joins with the Zamora Huayco river and forms the Zamora river, which keeps its name until it merges with the Namangoza river to create the Santiago river, which, as indicated earlier, is one of the main affluents of the Amazon river (see Fig. 2.9).

Zamora Huayco river

Zamora Huayco river is made with the flow of the Leon Huayco (Mendieta), Pizarro, El Carmen and San Simón streams and other minor springs (see Fig. 2.10).

Among the sources supplying water for the city of Loja are the El Carmen and San Simón streams, where the water is taken with conventional catchments – that is, with a lateral grid.

2.2 Climate of Ecuador

Ecuador is located at the northwest of South America between the geographic coordinates 01°27′06″ N–05°00′56″ S and 75°11′49″ W–81°00′40″ W, bordering on Colombia in the north, Perú in the east and south and the Pacific Ocean in the west. The Andes cross Ecuador from the north to the south, establishing a weather divide. Therefore, Ecuador can be divided into three climatic regions, besides the Galápagos Islands (Fig. 2.11):

- Coast.
- Highland (Sierra).
- Amazon basin.

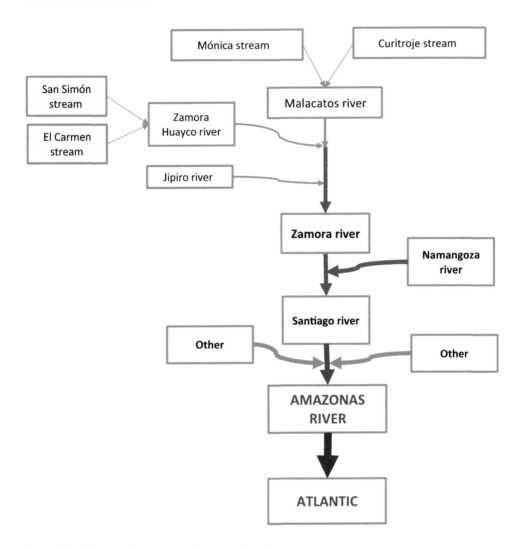

Figure 2.9 Abbreviated topological diagram of the Zamora river

Due to its position at the equator, Ecuador generally has a tropical climate, influenced by different atmospheric systems and ocean currents (Bendix and Lauer, 1992; Richter and Moreira-Muñoz, 2005). The main atmospheric systems are the Intertropical Convergence Zone (ITCZ; Equatorial Low), the Hadley Cell and anticyclones (Subtropical High) over the Pacific and the Atlantic, for which reason predominant wind direction is from the east (tropical easterlies, or trade winds; see Fig. 2.12a).

Additionally, two ocean currents influence the seasonal climate in Ecuador: the warm Pacific-Equatorial Counter Current, or Niño Current, and the cold Humboldt Current, or Perú Current, coming from the Antarctic (Fig. 2.12b).

Figure 2.10 Image of the junction of the San Simón and El Carmen streams, Loja – Ecuador

Figure 2.11 Climatic regions of Ecuador
Source: https://bit.ly/2zDHipN

(a)

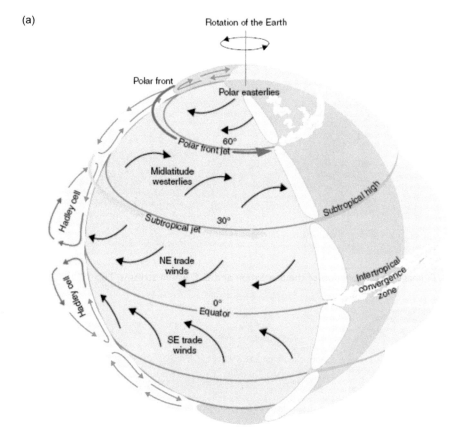

(b)

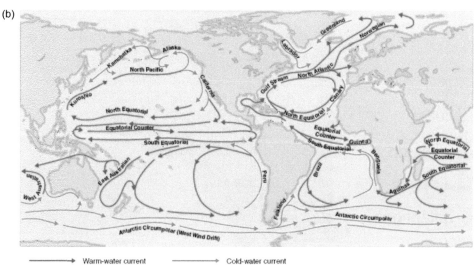

Warm-water current   Cold-water current

Figure 2.12 (a) Global atmospheric circulation, (b) global ocean currents
Source: https://bit.ly/2RH9NKH

The seasonal variations at the coast are determined mainly by the two ocean currents, establishing two annual seasons (rainy season and dry season). During austral summer (December–April), the Niño Current has greater influence on the coastal regions, which leads to higher amounts of precipitation, because the warmer ocean surface temperatures cause higher evaporation rates and, consequently, precipitation (rainy season).

In austral winter (June–September), the colder water of the Perú Current spreads out along the Ecuadorian coast, reducing evaporation rates and therefore precipitation amounts (dry season). May and October are transition months between the rainy season and the dry season (Rollenbeck and Bendix, 2011).

Mean annual precipitation at the coast is between 500 mm and 2000 mm, in which the northern parts are generally wetter, due to the longer influence of the warm Niño Current during the year. The rainfall extremes are observed in February/March, when evaporation rates are highest, due to vertical sun position near the equator (summit of austral summer). Mean annual temperature is around 26°C, with only small variations between the rainy season and the dry season (tropical climate; see Fig. 2.13).

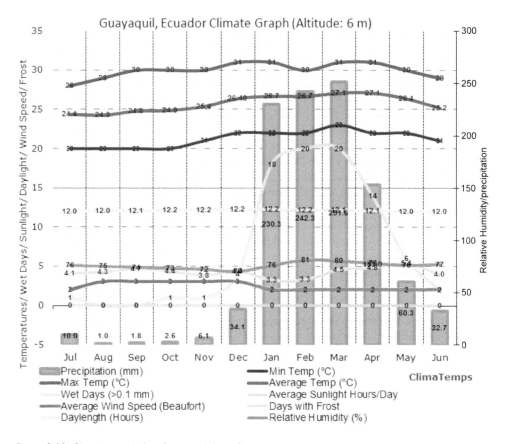

Figure 2.13 Climate graph for Guayaquil (coast)

Source: https://bit.ly/2PJBR2N

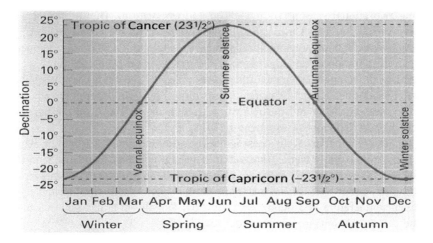

Figure 2.14 Time of vertical sun position at each latitude
Source: Strahler, 2013

The climate of the highland (Sierra) is related mainly to the sun position during the year. When sun position is vertical, more energy is received, which is why evaporation rates and precipitation amounts are higher. For Ecuador (1° north to 5° south), sun position is vertical in March/April and September/October (Fig. 2.14), when concurrently two precipitation extremes are observed.

However, the highland, in contrast to the coast, does not show a pronounced dry season, but climate can change rapidly between the valleys, which is caused by the local topography because individual mountain ridges form barriers to or pathways for humidity transport. In southern Ecuador, the two precipitation extremes are more weakly developed, because the times of vertical sun position become more approximate, for which reason only one large rainy season develops, between October and May (Oñate-Valdivieso *et al.*, 2018).

As mentioned before, annual precipitation amounts depend on the local topography, and vary generally between 700 mm and 1200 mm where bigger towns are built. Annual mean temperature depends on elevation, reaching temperatures clearly below 0°C at the highest mountain peaks, but where settlements exist, mean annual temperature lies between 10°C and 16°C (Fig. 2.15).

The climate in the Amazon basin is characterized by abundant rainfall throughout the year (humid-equatorial). However, this climate region presents two different rainfall regimes, specifically for areas below 1000 m asl (meters above sea level) and over 1000 m asl (Richter and Moreira-Muñoz, 2005). Regions below 1000 m asl show two precipitation extremes concurrently with vertical sun position. All months are humid, and mean annual precipitations varies between 2000 mm and 4000 mm, whereas the mean annual temperature ranges between 20°C and 22°C (Fig. 2.16). Regions over 1000 m asl show only one precipitation extreme during the year, strictly in June/July, when the tropical easterlies are most intense. The higher wind speed during austral winter makes the humidity transport from the Amazon basin most effective, provoking the precipitation extreme at the eastern escarpment of the Andes (over 1000 m asl).

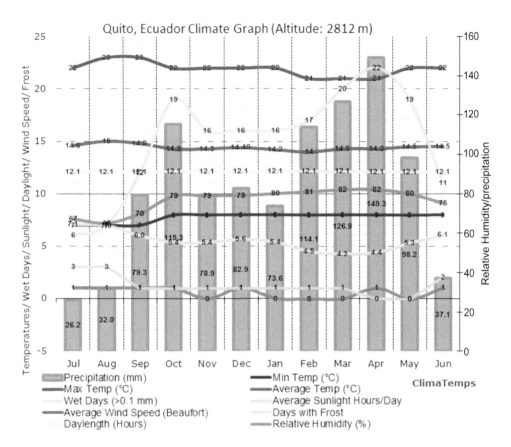

Figure 2.15 Climate graph for Quito (Sierra)

Source: https://bit.ly/2PM2dkp

Like the lower parts of the Amazon basin, all months are humid at the eastern escarpment of the Andes, but mean annual precipitation amounts and mean annual temperatures depend on local topography and on elevation (Fries *et al.*, 2014).

2.3   Climate of Loja

Loja is situated in an intra-Andean basin next to the eastern escarpment of the Andes in southern Ecuador between the geographic coordinates 3°51′47.58″ S–04°05′45.29″ S, and 79°07′58.26″ W–79°16′22.47″ W. The altitudes range from 1900 m asl at the river outlet (north) to 3400 m asl at the highest mountain tops (southeast; see Fig. 2.17).

For the analysis of climate conditions in the basin, information of eight automatic meteorological stations were used (Table 2.1). Five stations belong to the Technical University of Loja (Universidad Técnica Particular de Loja, UTPL; e.g. Oñate-Valdivieso *et al.*, 2018), which have been operating since 2011. Two stations are installed at the upper eastern ridge, operated and maintained since 1998 by investigation groups founded by the German Research Foundation (*Deutsche Forschungsgemeinschaft*, DFG; Tiro and Cajanuma; e.g.

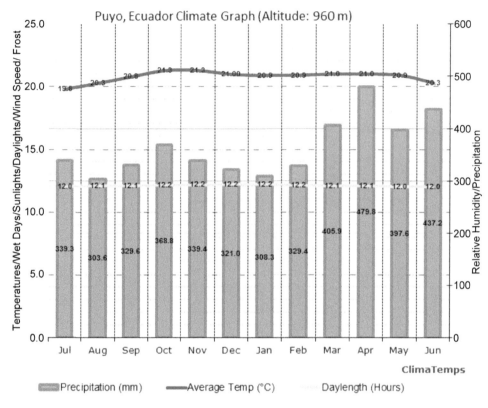

Figure 2.16 Climate graph for Puyo (Amazon basin)

Source: https://bit.ly/2SVNsuh

Ochoa-Cueva *et al.*, 2015). The station Argelia is the official weather station of Loja and belongs to the Ecuadorian weather service (*Instituto Nacional de Meteorología e Hidrología*, INAMHI), which provide historical information since 1965 (INAMHI, 2013). However, due to the different operation periods of the stations, thematical map generation was based on available information from 2011 to 2017.

Based on historical station data of Argelia (50 years), the climate in Loja can be classified as per-humid with a mean annual temperature of 16.0°C and a mean annual precipitation of 940.3 mm (Fig. 2.18). The rainy season occurs between December and April, whereas August and September are the driest months (Rollenbeck and Bendix, 2011; INAMHI, 2013). Wind directions are predominately from the east, due to the tropical easterlies (trade winds), carrying the humidity from the Amazon basin up to the eastern escarpment of the Andes. However, the eastern mountain ridge next to Loja forms a barrier, which is why monthly rainfall amounts decrease from the east to the west in the basin. Furthermore, precipitation amounts inside the basin depend on wind velocity and the humidity content of the air (moisture transport; Fries *et al.*, 2014).

In austral winter (June–September) easterly winds are strongest, reaching average monthly velocities of up to 15.5 m s^{-1}, but air humidity is lower, due to vertical sun position

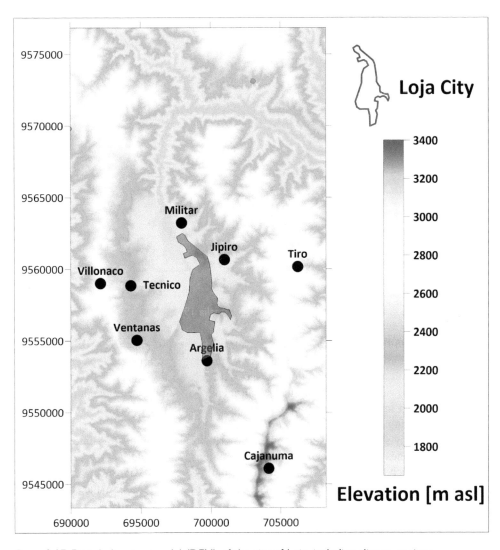

Figure 2.17 Digital elevation model (DEM) of the city of Loja, including climate stations

Table 2.1 Meteorological stations and available station data

Station	UTMX	UTMY	Altitude	Periods
Argelia	699711	9553629	2160	1965–2015
Militar	697901	9563240	2033	2011–2017
Jipiro	700975	9560679	2218	2011–2017
Tecnico	694294	9558872	2377	2011–2017
Ventanas	694716	9555060	2816	2011–2017
Villonaco	692138	9559012	2952	2011–2017
Tiro	706230	9560170	2850	1998–2015
Cajanuma	704140	9546097	3410	1998–2017

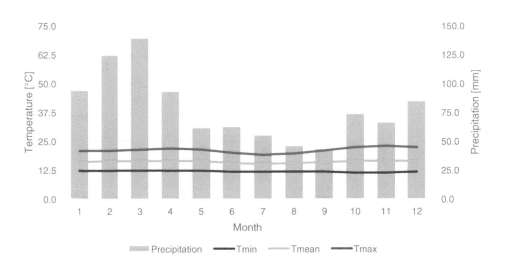

Figure 2.18 Climate graph for the city of Loja (station Argelia 1965–2015)

farthest away from the study area, precisely over the northern hemisphere (see Fig. 2.14), which reduces evaporation rates and the humidity content of the air (Windhorst *et al.*, 2013; Oñate-Valdivieso *et al.*, 2018). Additionally, most of the humidity content in the air precipitate is at the windward side of the eastern escarpment of the Andes (barrier), impeding the moisture transport into the valley, provoking only long-lasting drizzle with low precipitation amounts during this period (Fries *et al.*, 2014).

During the rainy season in austral summer, easterly winds are frequently interrupted by weak westerly winds from the Pacific Ocean, which bring generally dry and sunny weather to Loja (Rollenbeck and Bendix, 2011). However, during westerly winds, the formation of convective storms is frequent over the inter-Andean basins, provoking the main rainy season for these regions (Vuille *et al.*, 2000).

Due to the fast-changing topography in the basin of Loja (see Fig. 2.17), marked attitudinal gradients for air temperature, air humidity and rainfall are observed (e.g. Bendix *et al.*, 2009; Fries *et al.*, 2014). For a better understanding of the climate conditions in Loja, the following analysis describes only extreme months and the annual distribution of different climatic variables, based on daily station data from 2011 to 2017.

2.4 Solar radiation

The solar radiation depends principally on sun position, altitude and cloudiness (Barry and Chorley, 2010; Strahler, 2013). In general, during times of vertical sun position, more energy is received by the earth's surface, because the path through the atmosphere is shorter and, therefore, less energy is absorbed or reflected by gases and aerosols (particles). The same is valid for location at higher elevations, because the path through the atmosphere is reduced, which is why generally more energy is received at mountain peaks. Therefore, solar radiation shows generally positive altitudinal gradients; i.e. values increase with height. However, the exposition of slopes is important too, because slopes facing the sun receive more energy (see Fig. 2.14). This means for the equatorial zone, where Ecuador is located, that slopes facing north get more energy between April and September (austral winter, when the sun is vertical over the northern hemisphere) and slopes facing south get more energy between October and March (austral summer, when the sun is vertical over the southern hemisphere).

Nonetheless, the received energy is also linked to cloudiness, because a cloud cover absorbs and reflects big parts of the incoming solar radiation and only diffuse radiation reaches the surface. This notably reduces the incoming solar radiation. Therefore, if the sun's position is similar, more energy is received during dry seasons (fewer clouds) than during wet seasons.

In the basin of Loja, the situation is different, due to the observed high cloudiness at the mountain ridges, especially at the eastern barrier (Bendix *et al.*, 2004; González-Jaramillo *et al.*, 2016). The frequent cloud cover at the ridges provokes negative altitudinal gradients for the basin, which means that incoming solar radiation is higher inside the valley than at the mountain peaks (R^2: 0.75–0.87). Therefore, to create the monthly and annual radiation maps, kriging with detrended raw data was applied to illustrate the difference between the seasons (Fries *et al.*, 2009, 2012).

The months of extreme solar radiation in Loja are July (lowest) and November (highest). The low solar radiation in July is related to the sun position, which is farthest away from the basin (vertical over the northern hemisphere; see Fig. 2.14). Therefore, the path through the atmosphere is largest and the received solar energy lowest. Furthermore, during July, tropical easterlies are most intense, carrying the humidity from the Amazon basin up the eastern escarpment of the Andes, which leads to a constant cloud cover over Loja, especially at the ridges, producing long-lasting drizzle (Fries *et al.*, 2014). Figure 2.19a shows the solar radiation distribution in the basin during July, with maximum values at the valley bottom (137.6 W m^{-2}) and lowest at the mountain ridges (109.5 W m^{-2}), which is due to frequent cloud cover at the ridges.

The highest solar radiation is received in November, after the time of vertical sun position (September/October), because at the beginning of austral summer, tropical easterlies notably weaken and are frequently interrupted by periods of westerlies winds, which bring dry and

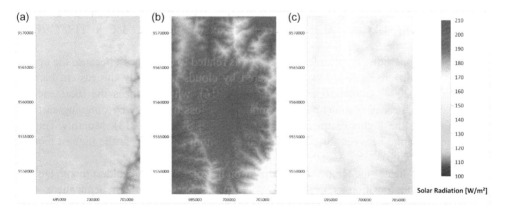

Figure 2.19 Average solar radiation for Loja: (a) July, (b) November and (c) year, based on information from 2011 to 2017

sunny weather to Loja (Vuille *et al.*, 2000; Rollenbeck and Bendix, 2011). Therefore, cloud-iness is reduced during November and longer dry periods exist, called by the local population "Veranillo del Niño" – the transition month between austral winter (dry season) and austral summer (wet season). Figure 2.19b illustrates the distribution of solar radiation in November, with the highest values at the valley bottom (208.8 W m^{-2}) and decreasing with height (157.7 W m^{-2}), which is due to the still-frequent cloud cover at the ridges (Bendix *et al.*, 2004).

Mean annual solar radiation (Fig. 2.19c) shows the same distribution as the extreme months, with values up to 172.7 W m^{-2} at the valley bottom and 124.8 W m^{-2} at the highest mountain peaks.

2.5   Air temperature

Air temperature is linked to solar radiation, because during times of vertical sun position, more energy is received and absorbed by the surface, causing higher air temperatures. Addi-tionally, air temperature depends directly on elevation (linearly; R^2: 0.92–0.99); generally decreasing with height in the troposphere (negative altitudinal gradients; Fries *et al.*, 2009). The global mean altitudinal gradient for air temperature in the troposphere is – 0.65°C/100 m (Strahler, 2013), but the gradient is not constant, because it also depends on the moisture content of the air. In humid air, altitudinal gradients decrease, whereas in dry air, gradients increase (Fries *et al.*, 2009).

Due to the direct dependency on elevation, the maps for average minimum (Tmin), mean (Tmean) and maximum (Tmax) temperatures for extreme months and the year were gener-ated by applying kriging with detrended raw data (Fries *et al.*, 2012). In general, minimum temperatures are observed closely before sunrise and maximum temperatures in the after-noon, between 2:00pm and 4:00pm (Barry and Chorley, 2010).

The extreme months of air temperature for Loja are July (coldest) and November (warmest), which can be related to the received solar energy (see Section 2.4). The highest monthly temperatures are observed in November (beginning of austral summer), closely after vertical sun position when cloudiness and precipitation amounts are reduced (see Fig. 2.18). This permits more solar energy to be received by the surface (Veranillo del Niño), and temperatures are higher.

Average Tmin for November range between 15.0°C at the valley bottom and 5.7°C at the highest mountain peaks (Fig. 2.20a), whereas Tmax vary from 27.3°C to 11.0°C (Fig. 2.20c).

The lower temperature amplitude at the ridges is related to cloudiness, because the upper parts of the basin are more frequently covered by clouds, especially the eastern barrier (Bendix *et al.*, 2004; González-Jaramillo *et al.*, 2016), which reduces the solar energy received during the day (cooler) but also impedes the loss of energy by outgoing surface radiation during the nights (warmer; Fries *et al.*, 2009; Strahler, 2013). Monthly Tmean for November range between 20.1°C at the valley bottom and 7.5°C at the highest moun-tain tops (Fig. 2.20b).

Figure 2.21 indicates the temperature distribution in Loja during the coldest month (July), when vertical sun position is farthest away from the basin (northern hemisphere; see Fig. 2.14), which reduces the solar energy received at the surface (Barry and Chorley, 2010).

Nonetheless, monthly Tmin at the valley bottom is slightly higher compared to Novem-ber, because a frequent cloud cover reduces nocturnal outgoing radiation over the basin in

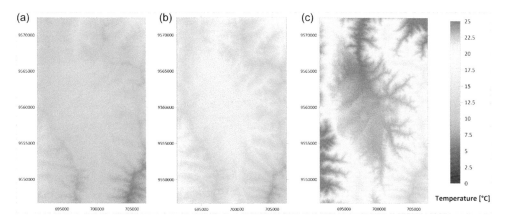

Figure 2.20 Monthly average minimum (a), mean (b) and maximum (c) temperatures for November, based on information from 2011 to 2017

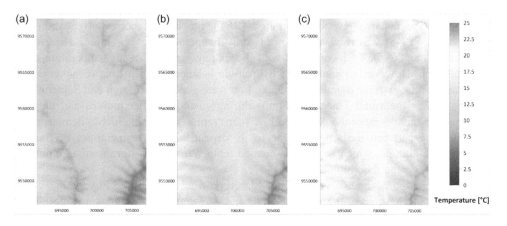

Figure 2.21 Monthly average minimum (a), mean (b) and maximum (c) temperatures for July, based on information from 2011 to 2017

austral winter (Tmin: 15.6°C). At the mountain ridges, Tmin in July (4.8°C) is lower compared to November (Fig. 2.21a), which is due mainly to the sun position. Tmax in July is notably lower compared to November, showing values from 22.7°C at the valley bottom to 7.1°C at the ridges (Fig. 2.21c). The same is valid for Tmean, displaying values between 18.5°C at the valley bottom and 5.6°C at the highest mountain peaks (Fig. 2.21b).

The small Tmean differences between austral winter and austral summer are due to the geographical position of Ecuador close to the equator, for which reason seasonal temperature variations are generally small, varying between 1°C and 2°C (Barry and Chorley, 2010; Strahler, 2013).

Average annual temperatures show the same horizontal and vertical distribution as the extreme months. The highest temperatures are observed at the valley bottom (Tmin: 15.7°C; Tmean: 19.5°C; Tmax: 24.9°C), whereas the lowest temperatures occur at the highest mountain ridges (Tmin: 5.6°C; Tmean: 6.7°C; Tmax: 8.9°C).

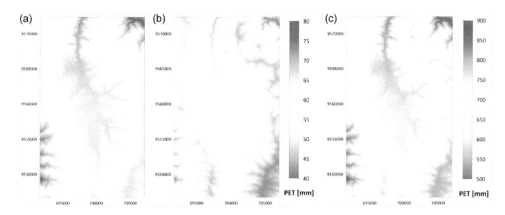

Figure 2.22 Average potential evapotranspiration for Loja: (a) March, (b) September and (c) year, based on information from 2011 to 2017

2.6 Potential evapotranspiration

To calculate the potential evapotranspiration (PET), the latitude corrected equation of Thornthwaite (1948) was used, which needs only temperature data as input. Therefore, the extreme months for PET coincided with the extreme months of air temperature and solar radiation. However, to analyze the hydrologic balance (HB) in Loja (Chapter 2.9), PET maps for the extreme months of precipitation were generated (March and September; see Fig. 2.18). The small Tmean variations during austral summer months and during austral winter months make the generated PET maps representative in order to demonstrate the difference between the seasons. Due to its dependence on temperature, PET also decrease with altitude, which is why the created temperature maps were used as input to calculate the average monthly and annual PET.

Figure 2.22a shows the average PET for March (austral summer; see Fig. 2.22a), with values between 76.9 mm at the valley bottom and 48.6 mm at the mountain ridges. The lower temperature in September (austral winter; see Fig. 2.22b) provoke lower evaporation rates, and PET is slightly lower, between 73.1 mm (valley bottom) and 42.0 mm (ridges). The small PET differences between austral summer and austral winter are due to the small temperature variations caused by the position of Loja close to the equator, where sun position is always high (Barry and Chorley, 2010; Strahler, 2013; see Fig. 2.14).

Average annual PET (period 2011–2017; Fig. 2.22c) is the sum of the individual mean monthly PETs, which ranges from 883.5 mm at the valley bottom to 546.7 mm at the highest mountain peaks. The lower PET at the ridges is caused by low temperatures and high cloud frequency (Bendix *et al.*, 2004), which additionally reduces evaporation rates because air is often close to saturation (RH ≤ 100%; Fries *et al.*, 2012; see Section 2.7). In general, evaporation rates depend on differences between actual relative humidity (RH) and air saturation (RH = 100%), which is why PET is reduced at higher elevations (Strahler, 2013).

2.7 Relative humidity

Air humidity depends on elevation, but in contrast to air temperature, RH shows an exponential behavior in relation to altitude. However, given only the data between 1900 m asl and 3400 m asl, the relation is nearly linear (Fries *et al.*, 2012), for which reason linear gradients were applied to generate monthly humidity maps (R^2: 0.85–0.98). Furthermore, in contrast to air temperature, the altitudinal gradients of RH are generally positive, which means that humidity increases with altitude (higher cloud frequency). Additionally, RH depends on temperature, because warm air can hold bigger amounts of water than cold air. Therefore, RH shows a clear daily cycle, with maximum values during the night (concurrently with Tmin) and minimum values during the afternoon (concurrently with Tmax) (Barry and Chorley, 2010).

The months of extreme RH in Loja are February (highest) and September (lowest). February is the summit of austral summer, when the sun position is nearly vertical over the basin, leading to the highest evaporation rates. Furthermore, Tmean is highest (February: Tmean = 19.5°C–7.1°C), which increases the capacity of the air to hold water, for which reason the concurrently highest precipitation amounts are observed (see Fig. 2.8). In September, during austral winter, lowest RH values are measured, because evaporation rates are lower due to the sun position being the farthest away from the basin (see Section 2.4). Additionally, the tropical easterlies become weaker at the end of austral winter, and the humidity transport from the Amazon basin is less effective.

Figure 2.23 shows the distribution of minimum (RHmin), mean (RHmean) and maximum (RHmax) RH for February in the basin of Loja. The lowest RH values are observed in the afternoon, between 54.9% at the valley bottom and 93.6% at the mountain ridges (Fig. 2.23a), whereas RHmax values, occurring during the night are generally close to air saturation, varying between 92.8% and 99.5% (Fig. 2.23b). RHmean ranges from 78.0% at lower elevations to 98.0% at the mountain tops (Fig. 2.23c), which illustrates the high evaporation rates during austral summer and the high cloud frequency at the ridges.

Figure 2.24 shows the RH distribution in Loja during the driest month in austral winter. In September, the tropical easterlies become weaker, and humidity transport from the Amazon basin is less effective besides generally lower air temperatures, which reduce the evaporation rates. RHmin in the afternoon varies between 50.7% at the valley bottom and 89.6% at the mountain tops (Fig. 2.24a). Nonetheless, in September, RHmax is also close to saturation, especially at the ridges, reaching values between 82.0% and

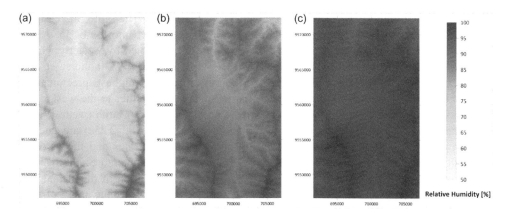

Figure 2.23 Monthly average minimum (a), mean (b) and maximum (c) RH for February, based on information from 2011 to 2017

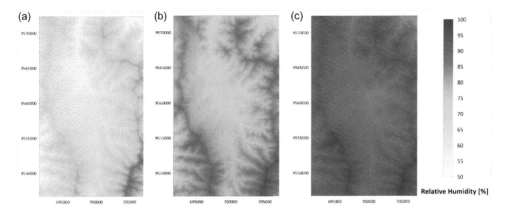

Figure 2.24 Monthly average minimum (a), mean (b) and maximum (c) RH for September, based on information from 2011 to 2017

99.0% (Fig. 2.24c; Fries *et al.*, 2012). Figure 2.24b shows RHmean for September, with values between 66.7% at valley bottom and 96.3% at the southeastern mountain peaks.

Average annual RH show the same distribution as the extreme months in the basin of Loja. Lowest RH are observed at the valley bottom (RHmin: 52.6%; RHmean: 73.3%; RHmax: 89.0%), and highest RH at the mountain ridges (RHmin: 90.9%; RHmean: 96.5%; RHmax: 99.0%) due to the dependency of RH on temperature, keeping in mind the extreme cloud frequency at the ridges (Fries *et al.*, 2012).

2.8   Precipitation

Precipitation amounts in mountain areas depend on moisture transport, differential surface warming, synoptic wind fields and local topography (Foresti and Pozdnoukhov, 2012). Mountain ridges in particular modify the spatial rainfall distribution, forming barriers or pathways for the moisture transport (Johansson and Chen, 2005). As mentioned before, the prevailing wind direction in the basin is from the east, due to the tropical easterlies, which carry the humidity up the eastern escarpment of the Andes. Therefore, the eastern ridge next to Loja forms a barrier, impeding the moisture transport into the basin. Also, during austral summer, when easterly winds are frequently interrupted by westerly winds from the Pacific, the proportion of the easterlies is still over 50% (Windhorst *et al.*, 2013; Oñate-Valdivieso *et al.*, 2018). As a consequence, the highest precipitation amounts can be expected at the upper eastern barrier, decreasing farther to the west.

As shown in Figure 2.18, the months of extreme precipitation are March (wettest) and September (driest). The highest precipitation amounts in March (Fig. 2.25a) are due to the high moisture content of the air, caused by high evaporation rates and the vertical sun position (see Fig. 2.14), which also leads to the highest temperatures during the summit of austral summer. Therefore, during this period, the formation of thunderstorms is frequent, especially during westerly wind conditions, which contribute most of the observed precipitation amounts (up to 80%) during this season (Rollenbeck and Bendix, 2011; Fries *et al.*, 2014; Fries *et al.*, 2020).

Highest precipitation amounts in March are displayed at the upper southeastern mountain ridge (up to 258.9 mm) decreasing farther to the northwest because of the

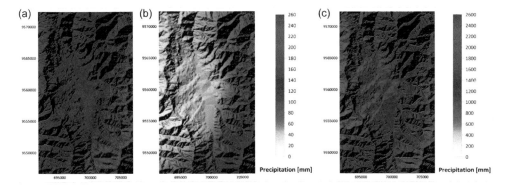

Figure 2.25 Average precipitation amounts in Loja: (a) March, (b) September and (c) year, based on information from 2011 to 2017

predominant wind direction from southeast and the barrier effect of the eastern mountain chain, which impede the moisture transport farther to the west. However, also the western ridge shows slightly higher precipitation amounts (up to 148.7 mm) than the valley bottom, because after the eastern barrier, air descends and clouds evaporates, but at the western ridge, air is forced to ascend again, and the remaining moisture content of the air condenses, causing new clouds to form and then rainfall (orographic rainfall; Foresti and Pozdnoukhov, 2012). Furthermore, during periods of westerly winds, thunderstorms are frequently formed over the inter-Andean basins, as over the neighboring basin of Catamayo in the west. The thunderstorms formed over Catamayo generally enter the basin of Loja from the southwest (lower mountain top altitudes), which is why higher precipitation amounts are observed there (Oñate-Valdivieso *et al.*, 2018). Lowest precipitation amounts are displayed in the north (91.6 mm), because this sector is particularly shielded from humidity transport by the high ridges located in the east and the west (Fig. 2.25a).

September (Fig. 2.25b) shows a similar precipitation distribution for Loja as it does for March, but rainfall amounts are notably lower. During austral winter, easterly winds are strongest, but RH is lower (see Section 2.7). Furthermore, most of the moisture content of the air precipitate is at the eastern barrier, where highest values (up to 138.5 mm) are measured. However, the strong barrier effect of the eastern ridge causes a fast decrease in precipitation farther to the west, leading to an average monthly precipitation of only 14.3 mm at the upper western mountain ridge. The western ridge forms a secondary barrier under easterly wind conditions, for which reason rainfall is generally zero in the neighboring basin of Catamayo during austral winter. Westerly winds are absent during this period; therefore, the western mountain ridge does not receive additional precipitation amounts.

The annual average precipitation map (Fig. 2.25c) displays the same rainfall distribution as what the extreme months display, which is due to the predominant wind direction from east-southeast. The barrier effect of the eastern ridge is clearly visible, where highest annual precipitation amounts are observed (2515.1 mm). Beyond the barrier, rainfall rapidly decreases, reaching only 679.9 mm at the upper western mountain ridge, which illustrates the strong barrier effect of the eastern mountain chain, reducing annual precipitation amounts by approximately 2000 mm within a distance of 10 km to 15 km.

2.9   Hydrologic balance

The hydrologic balance (HB) was calculated by means of the monthly and annual precipitation and PET maps, in which the PET maps (March, September and year) were subtracted from the respective precipitation maps (Barry and Chorley, 2010; Fries *et al.*, 2020).

Figure 2.26a shows the HB for March, the wettest month, where no water deficit is displayed for the whole basin, which is due to the high precipitation amounts during the summit of austral summer. The highest water surplus is display at the upper southeastern ridge (+205.4 mm), because rainfall is highest but evaporation lowest there (see Section 2.6 and Section 2.8). Beyond the barrier, precipitation amounts decrease, but concurrently, PET increases due to the higher temperatures and less rainfall (northwest; +26.5 mm). However, the western ridge also displays water surplus (between +40.0 mm and +60 mm), supported by extra precipitation received from thunderstorms formed over the neighboring basin of Catamayo. The extra rainfall mostly precipitates at the southern parts of the western ridge due to lower top altitudes, which is why the water surpluses are higher there.

In September (Fig. 2.26b), the HB of the basin of Loja is divided in an eastern part of water surplus (up to +92.8 mm) and a western part of water deficit (up to −48.4 mm). This is due to the prevailing easterly wind direction during austral winter and the strong barrier effect of eastern ridge, which impedes the moisture transport farther to the west. With increasing distance to the barrier, precipitation rapidly deceases, and PET rates exceed the precipitation amounts. Negative values were calculated from the valley bottom and up the western mountain ridge.

The annual HB map (Fig. 2.26c) generally shows a regulated water balance for the basin, with no water deficit. The highest water surplus is displayed at the eastern barrier, where values up to +1903.3 mm were calculated, for which reason this area is most important for the water supply. Farther to the west and down to the valley bottom, the water surplus rapidly decreases to +100 mm, due to lower precipitation amounts and higher PET. At the western ridge, water surplus is still lower (+50 mm) due to the larger distance to the eastern barrier. However, water deficit inside the valley generally does not occur, but behind the western ridge, which forms a secondary barrier, the water deficit is obvious, reaching − 115.3 mm at the upper parts. Therefore, irrigation is inevitable in the neighboring basin of Catamayo (west) but only seasonally necessary in the basin of Loja (Fries *et al.*, 2020).

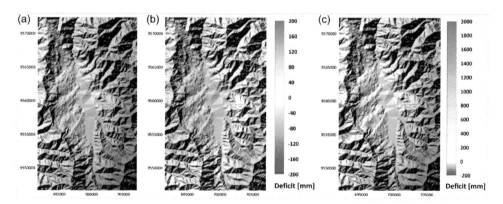

Figure 2.26 Average water surplus or water deficit for Loja: (a) March, (b) September and (c) year, based on information from 2011 to 2017

2.10 Case of hydrological study

The project site is located in the southeast of the Loja, because of the intersection of El Carmen and San Simón streams as a point of interest for the hydrological study. The mentioned micro-basins belong to the sub-basin of the Zamora river, located on the western slopes of the eastern range of the Andes. The natural drainage of these streams is oriented from southeast to northwest.

2.10.1 Study of the El Carmen and San Simón micro-basins

Here, we will list the factors that influence the characteristics of a hydrological system:

- Hydrological features of the basin: size, contour and surface, shape, average elevation, slope, drainage and topographic data.
- Aspects of the natural drains: disposition and length of streams, gradient thereof and the hydraulic components of the physical drainage channels network of the two micro-basins.
- River basin: the set of lands whose geomorphological features allow it to drain all the water (that is precipitated or accumulated) to a primary watercourse (river) and its affluents. The channel, of a river or stream, is the conduction of natural origin by which the water flows under the action of its weight in the direction of the bottom slope.

The project site is located southeast of the city of Loja, because of the intersection of the El Carmen and San Simón streams as a point of interest for the hydrological study.

The micro-basins of El Carmen and San Simón belong to the sub-basin of the Zamora river, located on the western slopes of the eastern range of the Andes. The natural drainage of these streams has an orientation from southeast to northwest.

2.10.2 Morphometric characteristics of the micro-basins

The characteristics of these micro-basins include size, shape, average elevation, slope and drainage.

Size of the basin

Geometrically, a basin is defined by two parameters: the *contour* that characterizes its shape and perimeter and the *surface* that it covers. It will quantify the totality of the drained area surface: from where the channel is born to the site of interest, comprising the perimeter of the basin. See Table 2.2 and Figures 2.27 and 2.28.

The El Carmen micro-basin encloses an area $Ac = 11.37$ km^2, and its contour includes a perimeter $Pc = 16.74$ km (Fig. 2.27 and Fig. 2.28).

Table 2.2 Perimeter and surface of the El Carmen and San Simón micro-basins

	El Carmen	San Simón
Basin perimeter	16.74 km	12.64 km
Basin area	1136.98 ha	629.76 ha

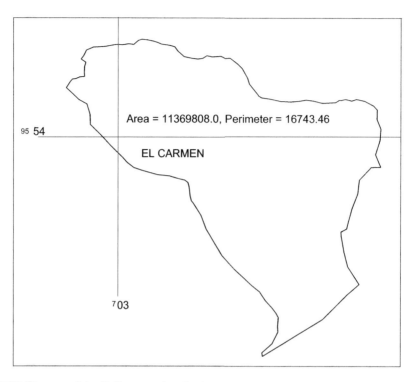

Area = 11369808.0, Perimeter = 16743.46

95 54

EL CARMEN

⁷03

Figure 2.27 Diagram of the El Carmen micro-basin

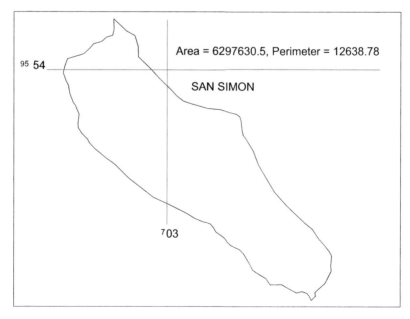

Area = 6297630.5, Perimeter = 12638.78

95 54

SAN SIMON

⁷03

Figure 2.28 Diagram of the San Simón micro-basin

Table 2.3 Tendencies in floods according to Kc

Basin shape	Compactness index Kc	Tendencies for floods
From almost round to round oval	From 1.00 to 1.25	High
From round oval to oblong-oval	From 1.25 to 1.50	Medium
From oblong-oval to rectangular	De 1.50 a 1.75	Low

Table 2.4 Gravelius coefficient (Kc)

	El Carmen	San Simón
Kc	1.40	1.42

The San Simón micro-basin encloses an area Ac = 6.30 km², and its contour includes a perimeter Pc = 12.64 km.

Shape of the basin

The amount of runoff from a micro-basin is a direct function of its form. This characteristic can be expressed with the compactness coefficient of Gravelius (Kc), which mathematically represents the ratio between the perimeter of the impluvium and the circumference that encloses an area equivalent to that of the drainage (Rai *et al.*, 2018).

$$Kc = \frac{Basin\,perimeter}{Circle\,perimeter}$$

$$Kc = \frac{Pc}{2\sqrt{\pi \cdot Ac}} \qquad\qquad [Eq.\ 2.1]$$

Description of variables:

Pc   basin perimeter, in km
Ac   basin area, in km²

The compactness index is related to the concentration time in the basin. If the value of Kc refers to the unity, then this time decreases and the risk of flooding increases. See Tables 2.3 and 2.4.

In conditions of an equivalent area for flat figures, the circle has the smallest perimeter, so that Kc will never be less than 1. So then, the closer Kc is to unity, the more the impluvium shape will be likened to a circle.

It can be inferred that both micro-basins have a round to oval-oblong shape, so it is expected that there is a common tendency for floods (see Tables 2.3 and 2.4). That is, the regime of streams is not stable but rather has a tendency for torrential rains.

Relief and average elevation of the basin

Surface runoff, infiltration, soil moisture and the contribution of groundwater to the flow of the stream are related to the slope of the basin (Tables 2.5, 2.6 and 2.7).

Table 2.5 Calculation of the mean slope of the El Carmen basin

Level curves (m asl)		Average curve		Area of each strip	Level difference	Sc
		Hi (m asl)	Li (km)	km² (Ai)	Di (km)	%
3420	3280	3350	1.09	0.13	0.140	1.338
3280	3200	3240	2.52	0.24	0.080	1.774
3200	3120	3160	3.52	0.44	0.080	2.473
3120	3040	3080	4.29	0.47	0.080	3.021
3040	2960	3000	4.50	0.47	0.080	3.163
2960	2880	2920	4.80	0.49	0.080	3.375
2880	2800	2840	5.74	0.60	0.080	4.041
2800	2720	2760	7.11	0.85	0.080	5.005
2720	2640	2680	7.99	1.10	0.080	5.623
2640	2560	2600	8.67	1.24	0.080	6.099
2560	2480	2520	9.38	1.42	0.080	6.596
2480	2400	2440	9.36	1.33	0.080	6.583
2400	2320	2360	8.48	1.31	0.080	5.964
2320	2240	2280	6.45	0.86	0.080	4.540
2240	2163	2202	2.52	0.42	0.077	1.705
			Total:	11.37	Slope:	61.30

Table 2.6 Calculation of the average slope of the San Simón basin

Level curves (m asl)		Average curve		Area of each strip	Level difference	Sc
		Hi (m asl)	Li (km)	km² (Ai)	Di (km)	%
3330	3280	3305	0.16	0.01	0.050	0.125
3280	3200	3240	0.44	0.05	0.080	0.554
3200	3120	3160	0.70	0.10	0.080	0.887
3120	3040	3080	0.96	0.11	0.080	1.218
3040	2960	3000	1.21	0.13	0.080	1.533
2960	2880	2920	1.59	0.18	0.080	2.015
2880	2800	2840	2.28	0.30	0.080	2.902
2800	2720	2760	3.20	0.42	0.080	4.059
2720	2640	2680	4.09	0.56	0.080	5.192
2640	2560	2600	4.88	0.82	0.080	6.205
2560	2480	2520	5.44	0.89	0.080	6.906
2480	2400	2440	5.60	0.88	0.080	7.117
2400	2320	2360	5.21	0.82	0.080	6.614
2320	2240	2280	3.97	0.74	0.080	5.042
2240	2163	2202	1.56	0.29	0.077	1.905
			Total:	6.30	Slope:	52.27

Table 2.7 Average percentage slope (Sc)

	El Carmen	San Simón
Sc	61.30	52.27

Also called decline, the slope affects the relationship between rain and runoff from the basin; therefore, the higher the hill, the faster the flow on the ground, the shorter the infiltration period and the shorter concentration time. The slope of the basin is calculated by the Horton–Alvord method. See Equation 2.2.

$$Sc = \frac{D_1 \cdot l_1}{Ac} + \frac{D}{Ac}(l_2 + l_3 + l_4 + ... + l_{n-1}) + \frac{D_n \cdot l_n}{Ac} \qquad \text{[Eq. 2.2]}$$

Specific slope

The slope index (Ig) decreases as the surface increases. The valuation of the specific slope (Ds) was used to determine this linear influence (see Eqs. 2.3–2.8).

$$Ds = Ig\sqrt{Ac} \qquad \text{[Eq. 2.3]}$$

$$Ig = \frac{H_5 - H_{95}}{Lm} \qquad \text{[Eq. 2.4]}$$

$$Lm = \frac{Pc}{4} + \sqrt{\frac{Pc^2}{16} - Ac} \qquad \text{[Eq. 2.5]}$$

$$Ln = \frac{Pc}{4} - \sqrt{\frac{Pc^2}{16} - Ac} \qquad \text{[Eq. 2.6]}$$

It must be fulfilled that:

$$Lm + Ln = \frac{P}{2} \qquad \text{[Eq. 2.7]}$$

$$Lm \times Ln = A \qquad \text{[Eq. 2.8]}$$

where
Ds   Specific gradient
Ig   Global slope index of P. Dubré, in m/km
Ac   Total area of the micro-basin under study in km²
H_5 Altitude at 5% of the hypsometric curve in m asl
H_{95} Altitude at 95% of the hypsometric curve in m asl
Lm   Greater side of the rectangle equivalent in km
Ln   Minor side of the rectangle equivalent in km
Pc   Perimeter of the basin in km

The value of the global slope index of P. Dubré in m/km (136.57 m/km and 151.37 m/km, respectively) warns us that the micro-basins of El Carmen and San Simón are characterized by having a solid relief (Tables 2.8 and 2.9).

Table 2.8 Global slope index of P. Dubré, in m/km (Ig)

	El Carmen	San Simón
Ig	136.57	151.37

Table 2.9 Type of relief based on lg

Type of relief	Range of values for lg
Very strong relief	lg < 2 m/km
Weak relief	2 m/km < lg < 5 m/km
Weak to moderate relief	5 m/km < lg < 10 m/km
Moderate relief	10 m/km < lg < 20 m/km
Moderate to strong relief	20 m/km < lg < 50 m/km
Strong relief	50 m/km < lg < 100 m/km
Very strong relief	100 m/km < lg < 200 m/km
Extremely strong relief	200 m/km < lg

Table 2.10 Specific gradient (Ds)

	El Carmen	San Simón
Ds	460.38	379.85

Table 2.11 Type of relief, depending on the Ds

Type of relief	Value range Ds
Very weak relief	Ds < 10 m
Weak relief	10 m < Ds < 25 m
Weak to moderate relief	25 m < Ds < 50 m
Moderate relief	50 m < Ds < 100 m
Moderate to strong relief	100 m < Ds < 250 m
Strong relief	250 m < Ds < 500 m
Very strong relief	500 m < Ds < 1000 m
Extremely strong relief	1000 m < Ds < 2500 m

The value of Ds (460.51 m and 379.85 m, respectively, in Tables 2.10 and 2.11) warns us that the micro-basins of El Carmen and San Simón have a strong relief.

2.10.3 Drainage

Establishing the type and distribution of the natural channels that exist in a basin allows knowing the efficiency of drainage and the relationship between the performance of the drainage network with the type of soil and surface of the basin.

The indices that measure the drainage system of a basin are two:

1   The density of the riverbeds (Dr), in riverbeds/km² (Eq. 2.9 and Table 2.12)

$$Dr = \frac{N}{Ac}$$

[Eq. 2.9]

2   The drainage density (Dd), in km/km² (Eq. 2.10 and Table 2.13)

$$Dd = \frac{\sum L}{Ac}$$

[Eq. 2.10]

Table 2.12 Density of the network (Dr) in riverbeds/km²

	El Carmen	San Simón
Dr	0.352	0.476

Table 2.13 Density of drainage, in km/km²

Characteristics of the basin	Density of the drainage
Regularly drained	0.0 < Dd < 1.0
Normally drained	1.0 < Dd < 1.5
Well drained	1.5 < Dd < 3.0
Highly drained	3.0 < Dd

where

N   Number of total channels in the basin (perennial and intermittent)
Σ_L Total length of water courses (includes perennials and intermittent), in km
Ac   Surface of the basin to the point of interest, in km²

The Dd gives us an indication of the efficiency of drainage in the basin.

With the drainage densities obtained in km/km² (0.88 and 0.98, respectively), we can infer that the micro-basins are regularly drained. That is, the micro-basins under study have a slow hydrological response, with resistant lithologies and where surface runoff has better opportunities (measurements in filtration time – percolation) to recharge aquifers in the area.

Slope of the channel

To find this parameter, the profile of the river is drawn from its source to the abscissa of interest. The points of change of slopes are joined with straight lines. On the abscissa axis, the length of the river is represented, and on the axis of the ordinates, the heights or elevation difference are represented; the vertical and horizontal scales are different, for obvious reasons (see Eq. 2.11). The process is as follows:

1   Calculate the area under the longitudinal profile.
2   Divide the obtained area (m²) between the length of the river (m) to obtain a value h (m). To this, the minimum height is added, and it is drawn in the center of the length of the river, which is the pivot point.
3   From the value of the minimum level, for the calculation length of the river, draw a straight line through the pivot point, this being the average slope.
4   From this line, determine points of maximum and minimum, with which we can quantify the average slope:

$$Sr = \left(\frac{Pmax - Pmin}{L} \right) \times 100 \qquad \text{[Eq. 2.11]}$$

where

Sr   Average slope of the river, in %
Pmax   Maximum level defined by d, in m asl
Pmin   Minimum level (corresponding to the point of interest), in m asl
L   Length of the river (from birth to the point of interest), in m

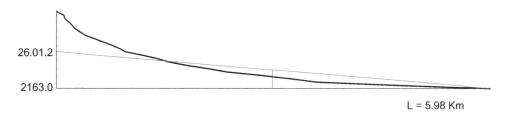

Figure 2.29 Longitudinal profile of the Quebrada El Carmen

Note: L = 5.98 km

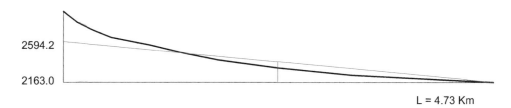

Figure 2.30 Longitudinal profile of the Quebrada San Simón

Note: L = 4.73 km

Length of the channel

The length of the main channel of the Quebrada El Carmen referred from its beginning to the point of importance is 5.98 km; in the same way for the Quebrada San Simón, we recorded a length of 4.73 km (Fig. 2.29 and Fig. 2.30).

$$Sr = \left(\frac{2601.2 - 2163}{5984.12} \right) \times 100$$

$$Sr = 7.54\% \qquad \qquad \ldots \text{quebrada El Carmen}$$

$$Sr = \left(\frac{2594.2 - 2163}{4729.3} \right) \times 100$$

$$Sr = 9.12\% \qquad \qquad \ldots \text{quebrada San Simón}$$

2.10.4   Average altitude

The average elevation of a basin is a factor that is related to temperature and precipitation; in turn, the temperature influences the evaporation. We will analyze two ways to determine it: a graphic one and a numerical one. See Tables 2.14 to 2.17 and Figures 2.31 and 2.32.

Table 2.14 Average elevation of El Carmen micro-basin (graphic method)

Level curves	Surface strip m²	km²	%	Altitude m asl	% accumulated areas	Length (m) of contour lines	Li (km) average
3420				3420	0	0	
3350	127527.61	0.13	1.14				1.087
3280				3280	1.14	2174.53	
3240	239228.92	0.24	2.11				2.522
3200				3200	3.25	2870.00	
3160	442574.92	0.44	3.87				3.515
3120				3120	7.12	4159.00	
3080	470869.56	0.47	4.13				4.293
3040				3040	11.26	4427.37	
3000	469272.85	0.47	4.13				4.496
2960				2960	15.39	4563.80	
2920	489477.28	0.49	4.31				4.797
2880				2880	19.70	5029.68	
2840	595255.92	0.60	5.28				5.743
2800				2800	24.98	6457.17	
2760	853637.97	0.85	7.48				7.114
2720				2720	32.45	7770.92	
2680	1100602.62	1.10	9.67				7.991
2640				2640	42.13	8212.05	
2600	1237163.67	1.24	10.91				8.668
2560				2560	53.03	9123.20	
2520	1416091.12	1.42	12.49				9.375
2480				2480	65.52	9627.50	
2440	1334756.08	1.33	11.70				9.356
2400				2400	77.22	9085.06	
2360	1312947.46	1.31	11.52				8.477
2320				2320	88.74	7868.40	
2280	861778.93	0.86	7.56				6.453
2240				2240	96.31	5036.60	
2201.5	418622.68	0.42	3.69				2.518
2163				2163	100.00	0.00	
Total area =		**11.37 km²**					

Precipitation

DETERMINING THE AVERAGE PRECIPITATION IN A BASIN

There are several methods to determine the average water sheet that falls in a basin by the availability of information. To calculate rainfall, the method of isohyet lines is adopted and contrasted with the precipitation gradient (see Table 2.18).

Equation 2.12 is used to determine the average precipitation of the basin (Pm), which is based on isohyet lines:

$$Pm = \frac{\sum_{i=1}^{n} P_i \times A_i}{A_t}$$

[Eq. 2.12]

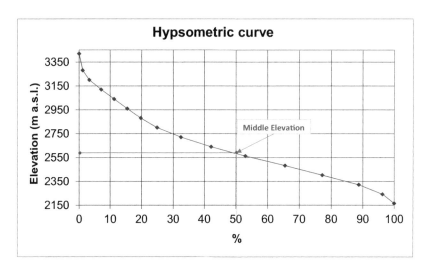

Figure 2.31 Hypsometric curve for El Carmen micro-basin

Table 2.15 Average elevation of El Carmen micro-basin (numerical method)

Level curves m asl		Average altitude m asl (Hi)	Surface strip km² (Ai)	Hi × Ai
3420	3280	3350	0.13	435.50
3280	3200	3240	0.24	777.60
3200	3120	3160	0.44	1390.40
3120	3040	3080	0.47	1447.60
3040	2960	3000	0.47	1410.00
2960	2880	2920	0.49	1430.80
2880	2800	2840	0.60	1704.00
2800	2720	2760	0.85	2346.00
2720	2640	2680	1.10	2948.00
2640	2560	2600	1.24	3224.00
2560	2480	2520	1.42	3578.40
2480	2400	2440	1.33	3245.20
2400	2320	2360	1.31	3091.60
2320	2240	2280	0.86	1960.80
2240	2163	2202	0.42	924.63
		Summation:	**11.37**	**29914.53**
		H average =	29914.53 11.37	
		H average =	**2631.01**	

where
Pi   Average rainfall between two consecutive isohyet lines
Ai   Area delimited by the two isohyet lines
At   Total area considered

For the average annual rainfall, a precipitation gradient was established between the stations: La Argelia, Cajanuma, San Francisco and Zamora. Because in the Zamora Huayco

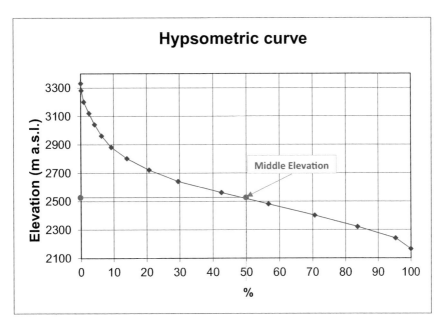

Figure 2.32 Hypsometric curve for San Simón micro-basin

micro-basin, located in the eastern part of the city of Loja, there are no weather stations, we used information from neighboring stations, the closest ones, and that will help in the development of the present hydrological analysis.

PRECIPITATION GRADIENT

The calculation of linear gradients can be applied because the horizontal distances are small and belong to the same tributary basin. The micro-basins of El Carmen and San Simón have, between the two, an area of 17.67 km², and the geomorphological characteristics are similar and very close to the La Argelia and Cajanuma stations that also belong to the same river basin. For these reasons, it is assumed that the linear relationship of precipitation and altitude is consistent; in addition, the correlation is within the parameter of very good. The stations of La Argelia and Cajanuma were taken to establish the linear rainfall gradient, with an information period of 35 years.

The equation of the line is a mathematical model that allows us to correlate precipitation and altitude for two stations that belong to the same tributary basin. The dependent variable is precipitation (mm), and the independent variable is the variation in altitude (m asl). See Equation 2.13, Figure 2.33 and Table 2.19.

$$Y = a + b \times X \qquad \text{[Eq. 2.13]}$$

where a and b are determined by least squares.

The precipitations generated by both methods are similar to each other. However, due to the more significant geomorphological relationship, due to the smaller distance between the

Table 2.16 Average elevation of the San Simón micro-basin (graphic method)

Level curves		Surface strip			Altitude m asl	% accumulated areas	Length (m) of contour lines	Li (km) average
	3330				3330	0	0	
3305		10717.03	0.01	0.17				0.157
	3280				3280	0.17	314.19	
3240		48140.89	0.05	0.76				0.436
	3200				3200	0.93	558.80	
3160		100644.74	0.10	1.60				0.698
	3120				3120	2.53	837.30	
3080		107616.71	0.11	1.71				0.959
	3040				3040	4.24	1081.50	
3000		134874.35	0.13	2.14				1.207
	2960				2960	6.38	1331.60	
2920		182186.42	0.18	2.89				1.586
	2880				2880	9.28	1840.75	
2840		301923.17	0.30	4.79				2.284
	2800				2800	14.07	2726.60	
2760		420740.57	0.42	6.68				3.195
	2720				2720	20.75	3664.07	
2680		555509.67	0.56	8.82				4.087
	2640				2640	29.57	4510.50	
2600		819529.3	0.82	13.01				4.884
	2560				2560	42.58	5258.20	
2520		890344.14	0.89	14.14				5.436
	2480				2480	56.72	5613.51	
2440		880917.32	0.88	13.99				5.612
	2400				2400	70.71	5589.91	
2360		816798.1	0.82	12.97				5.206
	2320				2320	83.68	4822.70	
2280		737474.27	0.74	11.71				3.969
	2240				2240	95.39	3115.67	
2201.5		290101.15	0.29	4.61				1.558
	2163				2163	100.00	0.00	
		Total area =	**6.30 km²**					

stations and the study basin and due to the topological similarities, we will adopt the precipitations found by the linear rainfall–altitude correlation method.

For the El Carmen micro-basin, an average annual precipitation of 1445 mm is expected, and for the San Simón micro-basin, an average annual precipitation of 1363 mm is expected.

WATER IN THE MICRO-BASINS

The water of El Carmen and San Simón streams is captured independently. However, they are joined downstream at the two catchments, and this flow is transported to the Pucará treatment plant in single adduction.

The water supply of these two streams constitutes 57% of the volume treated in the Pucará plant, which in turn supplies approximately 75% of the Loja population.

Table 2.17 Average elevation of the San Simón basin (numerical method)

Level curves m asl		Average altitude m asl (Hi)	Surface strip km² (Ai)	Hi × Ai
3330	3280	3305	0.01	35.36
3280	3200	3240	0.05	155.84
3200	3120	3160	0.10	317.90
3120	3040	3080	0.11	331.41
3040	2960	3000	0.13	404.70
2960	2880	2920	0.18	532.02
2880	2800	2840	0.30	857.40
2800	2720	2760	0.42	1161.13
2720	2640	2680	0.56	1488.74
2640	2560	2600	0.82	2130.70
2560	2480	2520	0.89	2243.56
2480	2400	2440	0.88	2149.40
2400	2320	2360	0.82	1927.65
2320	2240	2280	0.74	1681.50
2240	2163	2202	0.29	638.66
		Summation:	**6.30**	**16055.96**
		H average =	16055.96	
			6.30	
		H average =	**2549.66**	

Table 2.18 Generation of precipitations (isohyet lines)

Altitude strip	Precipitation (mm)	El Carmen area (m²)	Pi * Ai	San Simón area (m²)	Pi * Ai
1150					
1200	1175			217787.26	255900030.50
1250	1225	237572.69	291026545.25	571337.84	699888854.00
1300	1275	472632.25	602606118.75	611146.02	779211175.50
1350	1325	642129.87	850822077.75	565102.41	748760693.25
1400	1375	751390.12	1033161415.00	1339560.89	1841896223.75
1450	1425	1468343.87	2092390014.75	2316832.69	3301486583.25
1500	1475	2749910.46	4056117928.50	675859.41	996892629.75
1550	1525	2484601.99	3789018034.75		
1600	1575	1773391.44	2793091518.00		
1650	1625	789835.28	1283482330.00		
	SUMMATION	11369808	16791715982.75	6297626.52	8624036190.00
		Pm = 1476.87		**Pm = 1369.41**	

2.11   Hydrological-hydraulic study and risk of flooding in the Zamora Huayco basin

This chapter complements several studies carried out in the river basin, by means of a hydrological-hydraulic and flood analysis of the Zamora Huayco river basin, which includes the confluence of the San Simón and El Carmen basins, forming the channel that reaches the city of Loja, crossing it from southeast to northwest.

The main objective of this chapter is to obtain an approximation to the hydrological and hydraulic use of the basins studied, analyzing the maximum expected rainfall, taking into

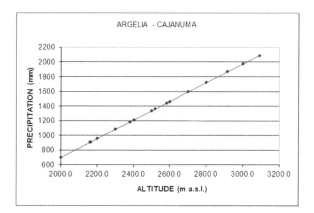

Figure 2.33 Precipitation–altitude relationship

Table 2.19 Summary of generated precipitations

Micro-basin	M. isohyets	M. correlation
El Carmen	1476.9	**1444.4**
San Simón	1396.4	**1362.8**

consideration a return period of 10 years and the way the existing network of channels drains such precipitation and looking at aspects such as the flooding area and the depth, the speed or the shear stress of the flow.

Finally, the problem of multiple risks is presented, in view of the specific case of soil erosion and flooding, since the former is considered a significant risk in the study.

2.11.1 Methodology

Scheme of the methodological process

To carry out the hydrological-hydraulic analysis and to obtain the flood area for a return period of 10 years, the starting point was set at the maximum daily precipitation of the basin investigated via a statistical analysis of the rainfall stations La Argelia-Loja, Cajanuma and Threw. The different flow rates have been derived from the aforementioned stations by the rationalized modified method by Témez.

The hydraulic study has been carried out by means of the two-dimensional modeling software HEC-RAS 5.0.5 as well as HEC-GeoRAS 10.1. The analysis process involves initial work that entails obtaining information and delineation. Then, hydrometeorological modeling is accomplished, with which the maximum instantaneous surface runoff flow for the aforementioned return period is calculated. Lastly, a hydraulic model is implemented to obtain the foreseeable flood sheet, including other data such as speed, depth or the shear stress of the flow.

The previous work took as the starting point a study of the basin and main channel. A delimitation of the basin is performed, and it is characterized, then the channel is digitized and a three-dimensional digital model of the land, at scale 1/3.000, is generated. At the same time, an inventory of all the crossing works, tubing, channeling, walls or bridges that may exist is done, taking into account their geometry and spatial location.

The next phase of the study is the hydrometeorological calculation of the basin, which consists of the calculation of the total daily precipitation and subsequently of the flow, by means of the rationalized method, modified according to Témez, as it is included in the 5–2-IC of Surface Drainage Standard of the Highway Instruction of the Ministerio de Fomento de España (Orden FOM/298/2016). Among all the existing hydrometeorological techniques, this is the one with the highest international acceptance, at least in accounts under 50 km^2, as is the case of those analyzed. As a result, in this phase, a design storm is generated, which with the characterization of the basin leads to the estimation of a flow of surface runoff in the drainage section investigated.

Finally, by using the two-dimensional hydraulic modeling software Hec-RAS 5.0.5, the behavior of the hydrographic network is established in the assumption of the stormy phenomenon estimated for a statistical period of 10 years. To achieve this, a series of transversal sections has been drawn up in the most representative sections of the channel and in each alteration of its profile. Moreover, the geometries of bridges, walls, pipes or any other anthropic or natural element that could influence the logical development of the flood flow are incorporated.

Summarizing the aforementioned parts of the study, the hydrological-hydraulic analysis of the basin includes the following information:

- A hydrological study of the basin and point of flow considered, for the return period of 10 years, including the following information:

  - Data of the basin (surface, length of the main channel, elevation of the head, elevation of the intersection, slope and average slope).
  - Rainfall (maximum daily rainfall expected for the considered return period).
  - Runoff coefficient.
  - Concentration time.
  - Average intensity of precipitation.
  - Flow (according to the rational method modified by J. R. Témez).

- Hydraulic detail analysis with HEC-RAS and HEC-GeoRAS for the study reach, graphing and/or detailing the following in tables:

  - Floodplain or ordinary maximum flood.
  - Roughness (Manning's coefficient).
  - Flow rates.
  - Water surface elevation.
  - Shear stress.

Characterization of the analyzed river basin

The point of flow considered for the study responds to a river basin of 2995 hectares, which includes the micro-basins of San Simón, El Carmen and Quebrada Medieta.

The basic data of its morphology are summarized as follows:

- Area: 29.95 km^2.
- Channel length: 8048 m.
- Header height: 3397 m.
- Intersection elevation: 2103 m.

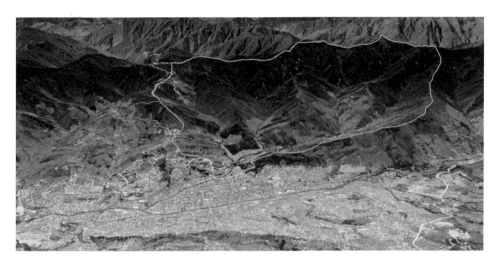

Figure 2.34 Three-dimensional representation of the drainage basin corresponding to the section
examined

Source: Google Earth

- Unevenness: 1294 m.
- Average slope: 16.1%.
- Time of concentration: 2.07 hours.
- Average intensity of precipitation: 5.33 mm h^{-1}.
- Runoff threshold (L m^{-2}): 25.4 mm.
- Runoff coefficient: 0.2.

2.11.2   Hydrometeorological calculation of the basin

Calculation of total daily precipitation

The total daily precipitation has been obtained through the statistical analysis of the daily
rainfall records of the La Argelia-Loja, Cajanuma and Tiro stations, located at the western,
northeastern and southern ends of the work basin, for the period 2006–2015.

   Taking as the starting point the average values for a return period of 10 years of the
aforementioned stations, an average value for the entire basin was obtained by means of
the inverse interpolation method of the weighted distance (IDW) in a geographic information system. This method has traditionally been used for the interpolation of precipitation
data, showing better accuracies than other similar methods (Cifuentes-Carvajal, 2016;
León-Gómez *et al.*, 2016). This value is 61.02 mm.

Calculation of the threshold and runoff coefficient

To obtain the flow of the watershed, the rational method provided in "Instruction 5.2-IC.
Surface drainage" of the Ministerio de Fomento de España (Orden FOM/298/2016) has
been applied. In the first place, a series of physiographic characteristics of the basin
have been calculated, such as its surface, maximum and minimum height, channel length

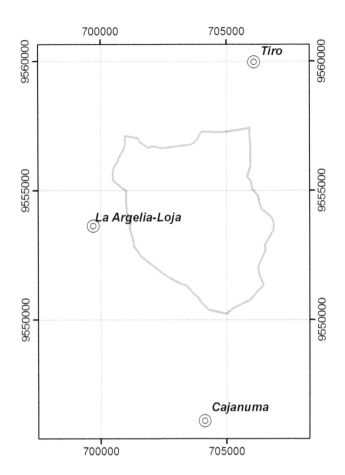

Station	UTM-X	UTM-Y	Altitude
La Argelia-Loja	699718.69	9553624.73	2160
Cajanuma	704138.09	9546097.33	3410
Tiro	706084.38	9559957.02	2814

Figure 2.35 Location of the stations used for rainfall analysis and UTM and altitude data

Source: From the author

and average slope. The time of concentration has been obtained by means of the following formulation (see Equation 2.14, time of concentration):

$$Tc = 0.3 \times \left[\left(\frac{L}{J^{1/4}} \right)^{0,76} \right]$$
[Eq. 2.14]

where
L   The length of the main channel (in km)
J   Its average slope (in m/m)

Furthermore, the average intensity of the precipitation (I_t) corresponding to the concentration time has been calculated. According to the formula in Equation 2.15, average intensity of precipitation is as follows:

$$\frac{I_t}{I_d} = \left(\frac{I_1}{I_d}\right)^{\left(\frac{28^{0.1}-t^{0.1}}{28^{0.1}-1}\right)}$$

[Eq. 2.15]

where

I_d (mm h^{-1})   The average daily precipitation intensity

I_1 (mm h^{-1})   The hourly intensity of the precipitation corresponding to the return period studied

t (h)   The previously calculated concentration time

These data are shown in Figure 2.36.

As a last step, the runoff coefficient has been determined. This coefficient defines the amount of rain that becomes real runoff, and it has been obtained from the runoff threshold, or, stated in another way, the amount of precipitation from which runoff begins. The value of the runoff threshold (Po) in a given basin, and for specific humidity conditions, is determined by the infiltration capacity of the soil, the use of land and agricultural activities and the slope of the land. This threshold has been calculated by using the curve number method of the United States Soil Conservation Service (USSCS, 1972). This method is widely accepted because of its capacity to estimate its parameters from soil and vegetation data. The Soil Conservation Service assumes the existence of a runoff threshold, Po, below which rainfall does not cause runoff. This value acts as an initial interception, before evaluating which part of it drains superficially and which part is retained.

Calculation of the flow

Once all the previous variables have been calculated, the flow has been obtained from the rational method, as indicated at the beginning of this section. This flow Q (m^3 s^{-1}) can be calculated according to the expression in Equation 2.16:

$$Q = \frac{C \cdot I(mm/h) \cdot A(km^2)}{3}$$

[Eq. 2.16]

where

C   The runoff coefficient

A   The area of the basin

I   The average intensity of precipitation corresponding to the considered return period and to an interval equal to the concentration time

This simple and universally accepted method is based on the transformation of a precipitation with intensity I, which starts instantaneously and continues indefinitely, to a runoff that will continue until the concentration time has been reached (Ct), at which time the whole basin is contributing to the flow. At that moment of equilibrium between inputs and outputs, the peak flow will be reached at the drainage point of the basin; the incoming volume will be the result of the intensity of precipitation by the area of the same (I × A),

CALCULATION OF FLOWS WITH THE MODIFIED RATIONAL METHOD

Basin: Zamora-Huayco (Loja, Ecuador)
Return period: 10 years

PLUVIOMETRY (Pd)
Total daily precipitation [mm]corresponding to the same return period

Pd =   **61.02**   mm

DATA OF THE BASIN

Surface (km2)	Length main channel (km)	Header height (km)	Elevation of the intersection (km)	Unevenness (km)	Medium slope (%)	Medium slope (m/m)
29.952	8.048	3.397	2.103	1.294	16.08	0.1608

RUNOFF COEFFICIENT (C)
Resulting runoff according to the daily rainfall Pd [mm] corresponding to the return period and the runoff threshold Po [mm]

$$C = \frac{[(P_d / P_O) - 1] \times [(P_d / P_O) + 23]}{[(P_d / P_O) + 11]^2}$$ C=   0.20

Runoff threshold Po (mm) =   25.0

CONCENTRATION TIME (Tc)
Concentration time [hours] for the study basin

$$Tc = 0.3 \times \left[\frac{L}{J^{1/4}}\right]^{0.76}$$ Tc =   2.07   hours

AVERAGE INTENSITY OF PRECIPITATION (It)
Average intensity of precipitation[mm/h] related to the concentration time

$$\frac{I_t}{I_d} = \left(\frac{I_1}{I_d}\right)^{\left(\frac{28^{0.1} - t^{0.1}}{28^{0.1} - 1}\right)}$$ It =   5.33   mm/h

.Id = Pd / 24 =   2.54   mm/h
I1 / Id =   2.50

FLOW (Q)
Maximum instant flow calculation [m3/s] according to the modified rational method by J.R. Témez

$$K = 1 + \frac{(Tc)^{1.25}}{14 + (Tc)^{1.25}}$$ K =   1.15

$$Q = \frac{C \cdot I \cdot A}{3,6} \cdot K$$ Q=   10.21

Figure 2.36 Flow calculation table according to the modified rational method
Source: From the author

and it is decreased by a runoff coefficient (C) between 0 and 1 that represents the proportion of water retained in the initial abstractions. In any case, the simplicity of this method is debatable, and its starting hypotheses (invariable precipitation along the basin in a Ct interval and constant runoff coefficient over time) are difficult to achieve in a natural system. To this end, a series of modifications have been suggested that aim to adapt it to rainfall with a longer duration than the time of concentration, larger basins and reconstructions, not only of the flow peak but also of the hydrograph, supposedly trapezoidal. The modification of Témez (Témez, 1978; Témez, 1991), shown in Equation 2.18, to apply to basins up to 3000 km^2 and concentration times between 0.25 and 24 hours, introduces a coefficient of uniformity of precipitation, K (see Equation 2.17), which can be calculated according to the concentration time.

Thus,

$$K = 1 + \frac{(Tc)^{1.25}}{14 + (Tc)^{1.25}}$$ [Eq. 2.17]

$$Q = \frac{C \cdot I \cdot A}{3} \cdot K$$ [Eq. 2.18]

Applying the rational method as just described results in a maximum instantaneous flow rate of 10.21 m^2 per second.

2.11.3   Hydraulic modeling

Once we know the flow for each basin or point of flow, the flood surface has been mapped according to the topographic characteristics of the land, the infrastructures and existing anthropic conditions, and so on.

The hydraulic calculation of the basin has been made using model HEC-RAS 5.0.5, consisting of the following:

- Tracing of control sections, bridges and sewers.
- Generating the channel geometry.
- Introducing hydraulic variables, such as Manning's roughness coefficient or the coefficients of contraction and expansion.
- Generating bridge geometry.
- Introducing flow data and boundary conditions.
- Calculating the sheet of water in each control section, along with other variables, such as flow rate, water surface elevation or shear stress.

To carry out the hydraulic modeling, it was necessary to create a hydraulic scheme of the channel and those infrastructures that act on it, either channeling it or altering its normal operation. This scheme consists of transverse sections based on the existing cartography for the study area, by means of a geographical information system.

The height of the water sheet for each control section and variables, such as flow velocity, wet section and so on, have been calculated by using the HEC-RAS model. Before starting this part, it is necessary to have the following georeferenced information:

- A layout of the channel.
- Transversal control sections. These sections have been drawn at those points where a change in flow behavior can be estimated, trying in any case that its distance does not exceed 50 m. In the profiles before and after bridges and casings and in the confluence of channels, coefficients of contraction and expansion have been introduced higher than normal in order to consider the narrowing imposed by the presence of the structure in the channel.
- Geometries of bridges and control sections of bridges. These sections represent the existence of an infrastructure that modifies the normal path of the flow.
- Bump mapping or the Manning coefficient for the basin. This value will depend on the use made of the soil, the existence of vegetation, the transversal location in the channel and so on.
- Flow data for the basin. This was already calculated in the previous section.
- A flow regime. Work has been done with a mixed regime, introducing critical boundary conditions both upstream and downstream, due to the existence of bridges that can modify a flow that in natural conditions would be fast or supercritical.
- A digital model of the field.

Results and discussion

A hydrological-hydraulic study on a stretch of channel 2536 m in length has been carried out, at the junction between the bank of the Zamora river and the city of Loja, collecting the immediately inland section, a rural area and the path of the river, which runs alongside the

Figure 2.37 Flooding sheet resulting in the neighborhood of the Zamora Huayco district and previous section

Source: From the author

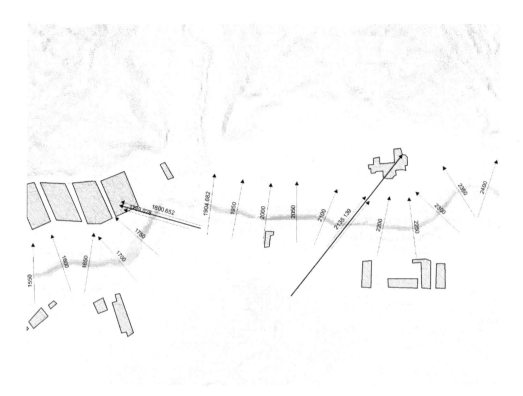

Figure 2.38 Depth of flow
Source: From the author

Zamora Huayco, Rodríguez Witt neighborhoods and also the surroundings of the Reina de El Cisne Federative Stadium. As a result of this analysis, it was determined that the channel is suitable for the period of return considered (ordinary floods). In the same way, the urban pattern of the section of the studied riverbed is appropriate for such flooding, since there is no intersection at any time with the flood flow. The entrance of the river into the city, which entails a narrowing of the channel and the appearance of bridges, significantly modifies or slows down the hydraulic behavior of the flow, generating a certain increase in depth that could lead to puddling on the roadway, though the heights of the flow would never exceed 20 cm.

Given the average values resulting from the section investigated, the flow velocity is 1.70 ms^{-1}, the flow area of the cross section is 8.87 m^2, the width of the floodplain is 11.34 m, the average height is 1.04 m and the cutting tension is 77.6 N m^{-2}.

Afterward, some of these values are cartographically represented for a specific section at the first connection between the Zamora river and the city of Loja.

Likewise, the water surface elevations and transversal profiles of the control sections of the studied part of channel are attached (see Fig. 2.41).

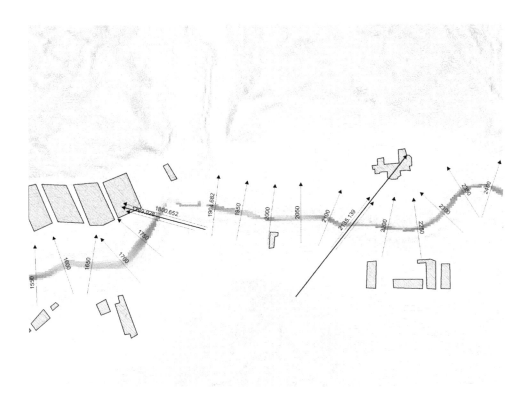

Figure 2.39 Flow speed
Source: From the author

Although floods in the study basin do not seem highly probable, the problem of erosion and how it acts in an associated manner, generating many risks, must be taken into account. As stated in Ochoa-Cueva and colleagues' investigation, in the same study region (2015), Ecuador has the highest rate of deforestation in South America. Between 2000 and 2010, 4 million hectares of forest land per year (FAO, 2010a) have been lost, and more than half of this area suffers from erosion problems. The Zamora Huayco basin, in particular, shows a high susceptibility to soil erosion due to pressure on forest cover due to anthropogenic activities. Following this conclusion, steeper slopes of agro-livestock use should be excluded (Ochoa-Cueva *et al.*, 2015).

The erosion of soils, the resulting contribution of sedimentation to the flow and eventually the level of grounding of the riverbed are regulated by the intensity of the downpour, the edaphological structure and texture, the protection of the soil against erosion, the slope or the tillage agricultural activities carried out in the headwaters. These are temporary variables with great impact on the magnitude of floods in the middle and lower zones and have the potential to become catastrophes. The degree of grounding of the channel explains, to a large extent, the depth and extension of the flood sheet, and it becomes a key indicator of

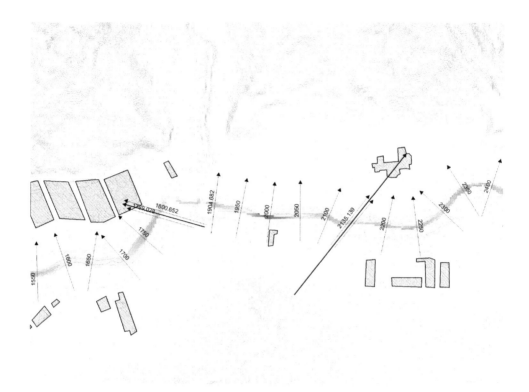

Figure 2.40 Cutting flow stress
Source: From the author

the stream and its basin health. Clean water avenues tend to evacuate without any problems, whereas catastrophic floods are frequently related to important solid flows (Gallegos Reina, 2015). In the middle and lower sections of the channels, when sedimenting, the riverbed is progressively raised, reducing the capacity of the drainage channel. In addition, the process feeds back, and in turn, the flood causes more erosion in some parts of the flood plain and in the river bed itself. Its widening and the undermining of the banks or the grooves are frequent phenomena during floods, which are also associated with a loss of soil productivity (Ayala-Carcedo and Olcina Cantos, 2002). Along with this, collapsing of the drainage system, hill sloping or the dismantling of infrastructure are likely (Fig. 2.42).

Some preventive measures that should be considered in this regard include (Gallegos Reina, 2015) incorporating the research regarding the potential soil erosion of the basin in the hydrological-hydraulic study, estimating the percentage of expected solid particles in the flood flow and determining the areas of erosion and the sedimentation in the basin. Likewise, in agricultural areas above a flooding area, plows should not be made too deep or parallel to the flow direction. Moreover, hydraulic measures such as retention dams, which prevent sudden runoff and consequent soil drag, should be incorporated. As a result of this investigation, we conclude by saying that the possibility of reclassifying and relocating land uses out of urban planning should be considered when it is demonstrated that the resulting solid sediments exceed the ideal thresholds.

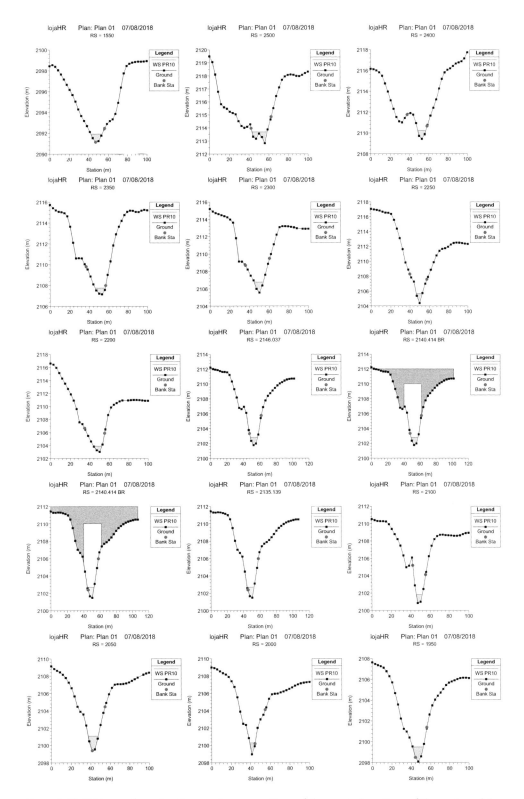

Figure 2.41 Water surface elevations and cross sections of the control sections of the sector of the studied channel

Source: From the author

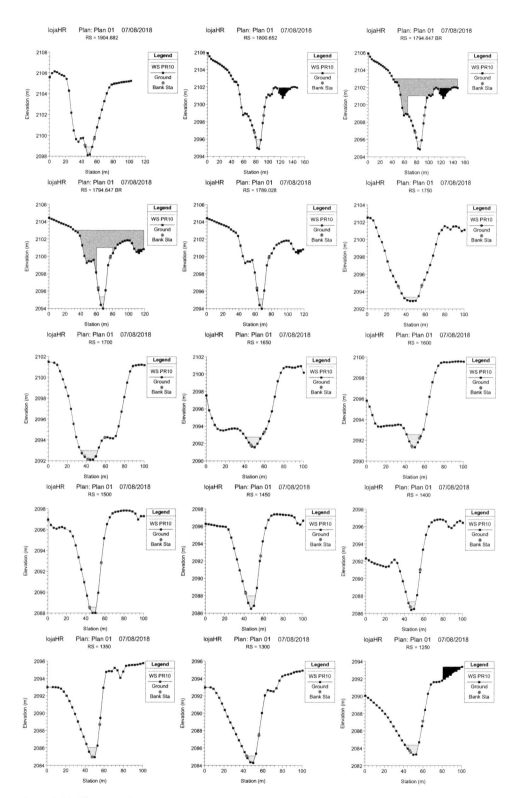

Figure 2.41 (Continued)

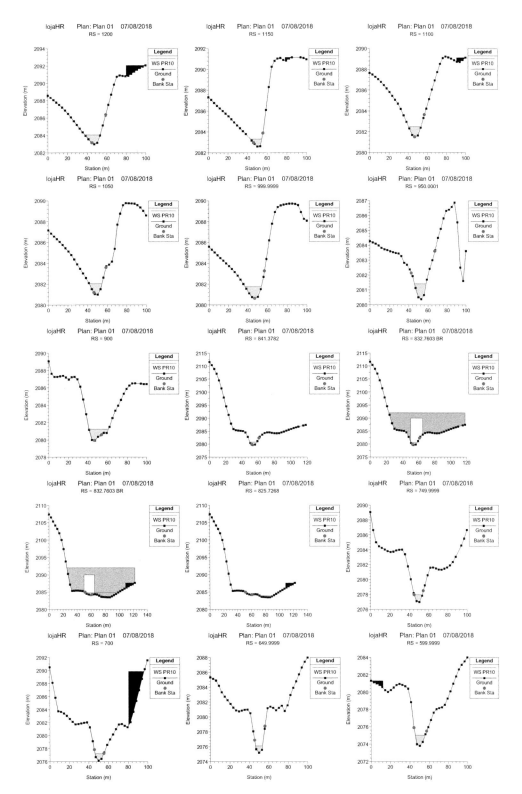

Figure 2.41 (Continued)

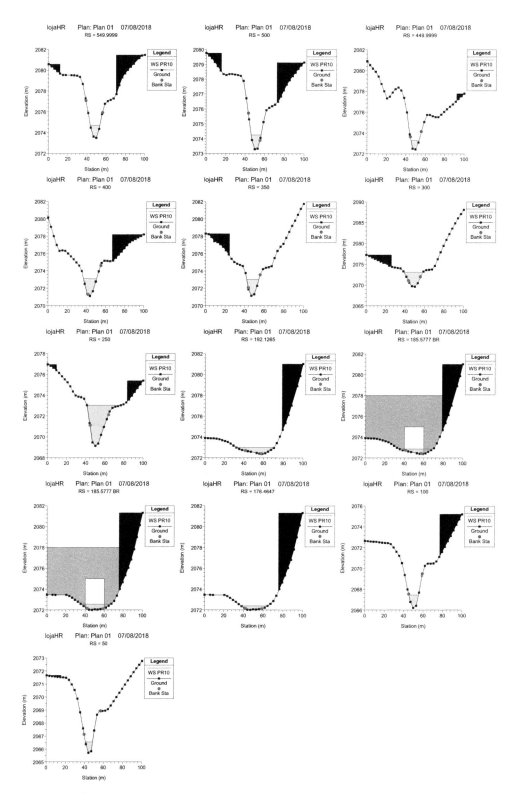

Figure 2.41 (Continued)

Figure 2.42 Accumulation of heterometric sediments and grounding in works of the passage of a stream

Source: From the author

2.12   Soil erosion

2.12.1   Generalities

Soil erosion is the major risk for watersheds worldwide, which is particularly evident in catchments affected by humans, where deforestation, road construction or building recreational facilities occur (Gorrab *et al.*, 2017). In Ecuador, the inter-Andean watersheds are especially affected, mainly due to land use changes caused by population growth (e.g. Molina *et al.*, 2007), which not only has consequences for water availability but also limits agricultural productivity (Du *et al.*, 2012; Fries *et al.*, 2020). Over deforested areas, the impact of rain drops (splash) is greater, which cause the sealing, slacking or crusting of soils, which leads to surface runoff increases and, consequently, to enhanced potential soil erosion, especially on steep slopes. Therefore, water availability decreases because less water can be stored in the ecosystem (Ochoa-Cueva *et al.*, 2015).

To estimate the potential soil erosion, this study uses the revised universal soil loss equation (RUSLE; Renard *et al.*, 1997), which is the most extensively used worldwide because it includes topographic, rainfall, soil and land-use data. Another advantage of RUSLE is that soil erosion is assessable also in watersheds without meteorological station data; only general information related to the watershed characteristics and local climatic conditions are necessary (Andrade *et al.*, 2010). Therefore, to illustrate the potential soil erosion in Loja, RUSLE was applied, and the Zamora Huayco watershed was selected, a sub-catchment in the southeast, where precipitation gradients are most extreme, but it provides drinking water for the local population and is used for recreational activities.

2.12.2   Characteristics of the Zamora Huayco watershed

The Zamora Huayco watershed is located at the southeastern flank inside the basin of Loja and verges on the buffer zone of the National Park Podocarpus in the east and south and on suburbs of Loja in the north and west. The catchment is situated between 3°59′24″ S and 4°03′8″ S, and 79°11′02″ W and 79°09′04″ W, covering an area of 37.3 km². The altitudes range from 2120 m asl at the valley bottom to 3420 m asl at the highest mountain tops (Fig. 2.43).

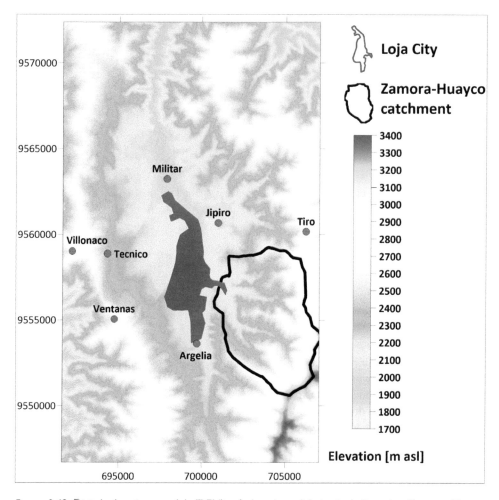

Figure 2.43 Digital elevation model (DEM) of the city of Loja, including the Zamora Huayco watershed

The main land cover units in the catchment are natural mountain forests, páramo, scrubland, pastures and urban area (Fig. 2.44). The anthropogenic land use changes are located mainly at valley bottom and in side valleys, where the natural vegetation was widely cleared to create pastures and agricultural land (Ochoa-Cueva *et al.*, 2015). At the middle and lower parts of the watershed, the predominant soils are Entisols, whereas at upper parts, the Inceptisols prevail (IEE, 2013). Soil texture can be classified principally in loam (53%) and clay loam (26%) (Martínez, 2009; Mejía-Veintimilla *et al.*, 2019).

2.12.3   Potential soil erosion in the Zamora Huayco watershed

The potential soil loss (A) was calculated by means of RUSLE (Renard *et al.*, 1997). The equation was executed for every grid cell (x,y) with a spatial resolution of 50 m × 50 m

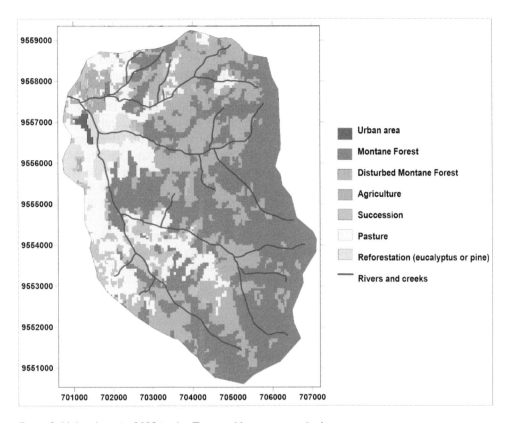

Figure 2.44 Land use in 2008 in the Zamora Huayco watershed

Source: Ochoa-Cueva *et al.*, 2015

(Ochoa-Cueva *et al.*, 2015):

$$A_{(x,y)} = R_{(x,y)}K_{(x,y)}LS_{(x,y)}C_{(x,y)}P_{(x,y)}$$ [Eq. 2.19]

where

$A_{(x,y)}$ The average soil loss produced by water erosion per unit of area [t ha^{-1} a^{-1}] for every grid cell (x,y)

$R_{(x,y)}$ The erosivity factor caused by rain [MJ mm ha^{-1} h^{-1} a^{-1}] for every grid cell (x,y)

$K_{(x,y)}$ The erodibility factor of the soil [t h MJ^{-1} mm^{-1}] for every grid cell (x,y)

$LS_{(x,y)}$ The length and degree of the slope for every grid cell (x,y)

$C_{(x,y)}$ The coverage and management factor for every grid cell (x,y)

$P_{(x,y)}$ The support practices for every grid cell (x,y)

To determine the individual factor at each grid cell (x,y), first, the *R*-Factor was calculated, using precipitation data from the three meteorological stations closest to the study catchment, specifically the official weather station of INAMHI (Argelia is 2160 m asl) and the two stations situated at the eastern mountain ridge (El Tiro is 2850 m asl and Cajanuma is 3410 m asl; see Fig. 2.17). However, only daily information was available, for which reason the R-Factor equation from Renard and Freimund (1994) was applied, which uses the modified Fournier index (MFI) based on average monthly and annual precipitation data (Ochoa-Cueva *et al.*, 2015). Then, the erodibility rate or rate of soil loss (K-Factor) was determined, which depends on soil characteristics; to this purpose, 38 soil samples were taken and analyzed. The results for each sample point were interpolated by kriging to obtain the respective K-Factor at each grid cell (x,y) (Pérez-Rodríguez *et al.*, 2007). Afterwards, the topographic variables (LS-Factor) were derived using a DEM with a 50 m × 50 m resolution (Fig. 2.43) to obtain slope length (L) and the slope steepness (S) at each grid cell (x,y), by applying the equation proposed by Mitasova *et al.* (1996) and Nearing (1997), respectively. To estimate the C-Factor at each grid cell (x,y), the vegetation map (Fig. 2.29) was used, which was generated by means of ASTER satellite images from 2008, applying a supervised classification (Richter, 2007). Finally, the P-Factor was set to 1, because no support practices to reduce soil erosion have been implemented in the Zamora Huayco watershed (Ochoa-Cueva *et al.*, 2015).

The calculated annual potential soil loss (A) is shown in Figure 2.45. Minimum values were calculated for areas with natural vegetation (primary mountain forest and páramo) or low slopes (1.5 t ha^{-1}), whereas maximum values were determined for agriculture land

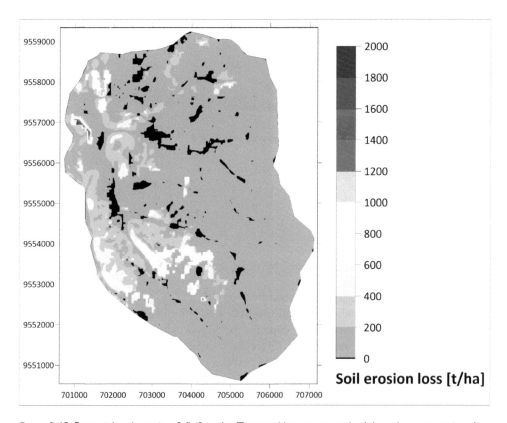

Figure 2.45 Potential soil erosion [t/ha] in the Zamora Huayco watershed, based on vegetation data from 2008

Source: Ochoa-Cueva *et al.*, 2015

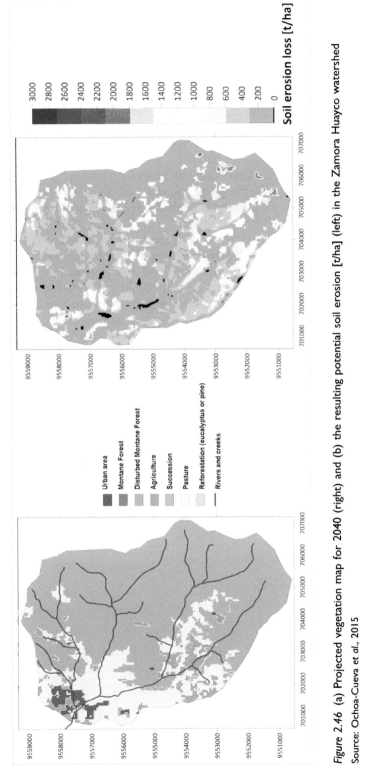

Figure 2.46 (a) Projected vegetation map for 2040 (right) and (b) the resulting potential soil erosion [t/ha] (left) in the Zamora Huayco watershed

Source: Ochoa-Cueva et al., 2015

and areas with steep slopes covered by pastures (1894.2 t ha^{-1}). The calculated erosion at the most vulnerable parts correspond to a loss of more than 7 cm of topsoil per year, which are mostly located in the northwest of the watershed, where the anthropogenic impact is obvious. Normally, the potential soil erosion should be reduced at these parts, because slope steepness is lower as well as precipitation amounts. However, the land use change at these areas left the soil widely unprotected, so soil erosion processes are intensified. The upper ridges have steeper slopes and concurrently show highest precipitation amounts; nonetheless, these areas show minimum values, because the soil is protected by natural vegetation, which also provides a high infiltration capacity (plant roots) that reduces surface runoff (Molina *et al.*, 2007).

In general, natural vegetation cover (tropical mountain forest and páramo) protects the soil from erosion processes, due the sheltering effect of the canopy layer (splash, Fries *et al.*, 2020). Furthermore, the plant roots retain the soils even at steeper slopes (Körner and Paulsen, 2004).

If land use change progresses as observed in the past, primary forests and páramo will disappear mid-century in the Zamora Huayco watershed (Ochoa-Cueva *et al.*, 2015). To illustrate these changes, a projection for the year 2040 was generated, applying the same deforestation rates as observed between 1976 and 2008 (Fig. 2.46). The hypothetical land use implies that most of the montane forest will be intervened, and only disturbed montane forest will remain (Fig. 2.46a). Urban areas, agriculture and pastureland will expand considerably, even on steeper slopes. This future scenario leads to a dramatic increase in potential soil erosion (up to 3040.9 t ha^{-1}; see Fig. 2.46b) assuming that the 2040 scenario is the worst case, because deforestation rates will hopefully decrease in the future.

This analysis identified vegetation cover (C-Factor) as one of the principal factors for soil erosion risk in the Zamora Huayco watershed. The general assumption that precipitation (R-Factor) and topography (LS-factors) have the greatest influence on soil erosion is valid only for homogeneous vegetation covers and topography. This can be illustrated by means of the southeastern mountain ridge, where climatic and topographic conditions are most extreme but lowest potential soil erosion was calculated, because under natural vegetation, soils are protected.

Chapter 3

Market study on drinking water

Holger Benavides-Muñoz, Jorge Arias-Zari and
José Sánchez-Paladines

Two market studies on drinking water were developed in the city of Loja, Ecuador, the first in 2005[1] and the other in 2011.[2]

3.1 Universe and samples in market studies

For the first market study in 2005, a universe of 24,410 users connected to the drinking water system in the city of Loja was adopted. The sample size was calculated for an error of 5% and a level of reliability of 95%, by using the expression in Equation 3.:

$$n = \frac{pqN}{Ne^2 + pq}$$ [Eq. 3.1]

where
n Sample size
N Universe (24,410)
p Percentage of favorable cases (0.5)
q Percentage of unfavorable cases (0.5)
e Error under a certain level of confidence (5%)

With the respective replacement for the calculation, it is obtained that n ≈ 100. Consequently, the sample size calculated for the 2005 study is 100.

For the 2011 market study, the sample considered was 100% of the users with potable water supply service in the urban neighborhoods La Tebaida, Bellavista and San Pedro in the city of Loja, because 1183 users with a domiciliary meter had to be investigated to update the cadaster of the hydrometric district to which they are connected, and they gave a positive response to the survey. Next, the comparison of the results of both studies will be made.

The limits of the study area include the following:

North: Alonso de Mercadillo Street
West: streets such as Panama, Guayana, Paraguay, Galápagos,
Argentina and the Dominican Republic
South: Argentina Street
East: Pío Jaramillo Alvarado Avenue

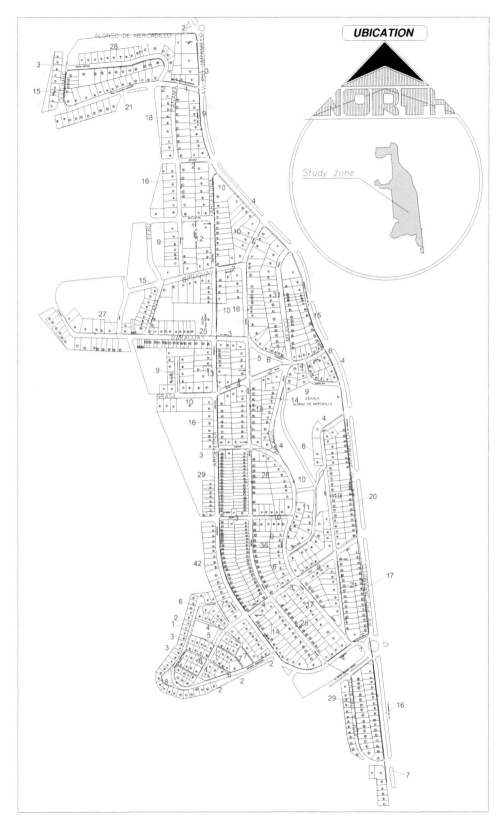

Figure 3.1 Hydraulic delimitation of La Tebaida hydrometric district

3.2   Characteristics of the surveys applied in market studies

The surveys were structured to determine, among other parameters, the availability of the respondents to increase the payment in the monthly bills for maintenance and conservation of the micro-tributary basins of El Carmen and San Simón. They also aimed to determine the opinion regarding the acquisition by UMAPAL of the land (farms) for reforesting and ensuring sustainability for them. The surveys had multiple-choice and open-type questions.

3.2.1   Market studies

The general data of the respondents are displayed graphically.

Study of Benavides-Muñoz, H. and Arias, J. (2005) and Study of Benavides-Muñoz, H. and Paltín, G. (2011).

Study 2005	Study 2011
About the number of people living in housing, 73% are located in the ranges from 4 to 6 individuals.	Of the properties surveyed, **93.83%** are inhabited by an average of 7 people per family per property.

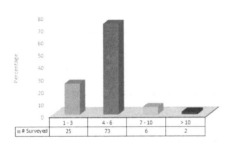

Figure 3.2 Number of people per dwelling

73% of respondents indicated that the number of people living in their home is between 4 and 6 people, 25% had fewer than 3 people and the remaining 8% indicated that more than 7 people live by the property. The arithmetic average is considered to be **4.5 people/housing**.

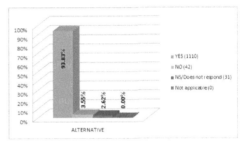

Figure 3.3 Inhabited premises

According to the general statistical average for the hydrometric district studied, **5.8 people are projected per property**; 3.55% are uninhabited, and 2.62% did not know or did not answer.

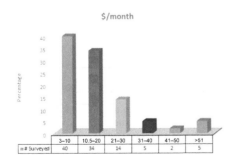

Figure 3.4 Monthly representation by water bills

40% of respondents pay between US$ 3 and US$ 10; 34% between US$ 10.5 and US$ 20; 14%

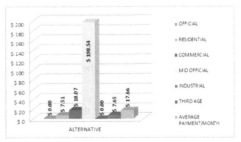

Figure 3.5 Monthly   payment   for   water consumption

Regarding the average monthly payment by categories, there are zero users with official rate,

(Continued)

(Continued)

Study 2005	Study 2011

pay between US$ 21 and US$ 30; 5% pay between US$ 31 and US$ 40; 2% pay between US$ 41 and US$ 50; and 5% per month pay over US$ 51.

The general average is US$ 19.6 per month per customer.

In the results of the respondents, the data were provided by the owners of residential houses and those with centric housing uses, for renting rooms and small apartments.

in the residential category. They paid an average of US$ 7.51 per month, commercial US$ 18.07, institutional US$ 198.54 and senior citizens US$ 7.65. In the hydrometric district, properties are not classified as industries.

The general average is US$ 17.66 per month per customer.

In the results of the respondents, the data were provided by those owners of houses where urban family housing predominates, with little rental housing.

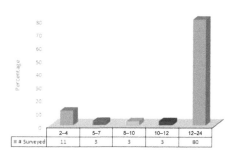

Figure 3.6 Time in hours of the available water service per day

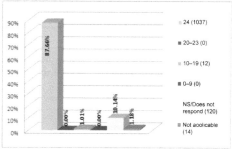

Figure 3.7 Hours of water service per day

Of the respondents, 11% have only up to 4 hours per day of drinking water service.

9% have water service for a daily period between 5 and 12 hours per day.

80% of the respondents answered that they had water between 12 and 24 hours per day.

About the hours of daily drinking water service provided by UMAPAL (2011), the results indicate that a total of 87.66% have water all day, 1.01% of users have service from 19 to 10 hours a day, 10.14% did not respond and the remaining 1.18% do not apply to this question.

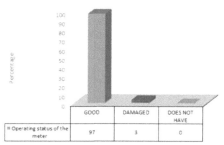

Figure 3.8 Characteristics of the counter

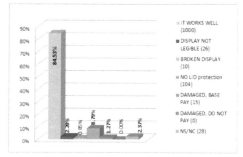

Figure 3.9 State of the counter

Source: Benavides-Muñoz and Paltín (2011)

97% of those surveyed stated that their home counter is in perfect condition; besides, 3% were able to indicate that it is damaged and that they pay the consumption base.

The counters observed are in the following conditions: 84.53% of them usually work and a good state of conservation is evident, 2.20% have an opaque moon, 0.85% have the moon broken, 8.79% have their lid absent, 1.27% of them are damaged but pay the base, 0.00% are damaged and do not pay for the service and 2.37% of users do not know or do not answer the question.

(Continued)

(Continued)

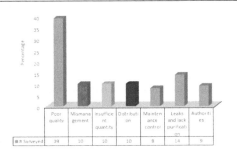

Figure 3.10 Problems perceived in the service

The 39% of respondents perceive as a problem the poor quality of the water sold; 19% assume that it is the lousy administration, the authorities and the population growth; and only 8% believe that the problem is control and maintenance.

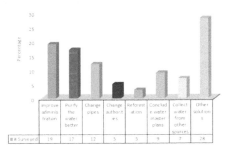

Figure 3.12 Solution to perceived problems

The respondents indicated that one of the most significant perceived problems is the poor water quality, and they consider that the solutions would be to improve the administration, improve the treatment in the plant and change the distribution pipes.

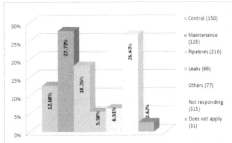

Figure 3.11 Perception of service problems

12.68% of users perceive that the lack of control is one of the problems in the drinking water service, another 27.73% indicate that the problems can be due to the lack of maintenance, 18.26% believe that the problem is the poor state of the pipes, 5.58% of respondents suggest that leaks cause severe problems in the service, 6.51% mention some other issues, 26.63% did not know or did not answer and 2.62% do not apply to this question.

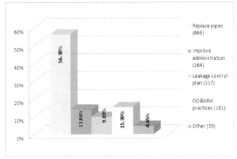

Figure 3.13 Proposed solution to problems

According to the respondents, the solutions to improve the potable water service include replacing pipes (56.30%), improving the administration (13.86%), implementing a leakage control plan through the Infrastructure Leakage Index (ILI) (García-Espinosa & Benavides-Muñoz, 2019) and avoiding wasting water (9.89%), improving the practices of operation and maintenance (OO & MM) (15.30%) and searching for other solutions (4.65%).

3.2.2 Calculating the value of the water resource

The principle of economic efficiency requires that water service rates be designated on the basis of the calculation of the opportunity cost of the goods and services involved. In competitive markets, the opportunity cost is equal to the short-term equilibrium price of these goods and services, as a result of free supply and demand. In this context, and always assuming that there is no benchmark rate value valid in the surrounding market, the recommended action to obtain such a value, with the maximum efficiency in the distribution of the resource, is simulating the equilibrium that would originate in a hypothetical competitive market in which the same service is negotiated.

In the present case of study, the economic value of water is estimated through the contingent valuation method (Zapata *et al.*, 2012), for which a *possible theoretical circumstance* is considered, such as to facilitate the determination of the maximum *payment availability* that the claimants would be determined to pay so long as the quality of the drinking water service improves in the near future.

Possible hypothetical circumstance

In both of the market studies that were carried out, the subscribers and users of the drinking water system and the voluntary amount of money that they would be willing to pay in addition in the monthly payroll for the collection of the water service, as a contribution to the improvement in the technical management of the system and the acquisition of the lands of the micro-watersheds from where they are supplied, were determined.

The comparative result of this question, in both of the market studies, is presented as follows:

Study 2005	Study 2011

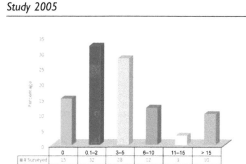 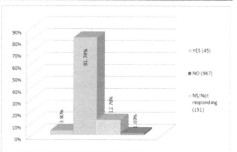

Figure 3.14 Incremental payment availability

On average, 85% of the respondents would be willing to pay voluntarily: US$ 5.86 per month (Zapata *et al.*, 2012). Then the consumer surplus is US$ 5.86 per month or US$ 70.32 per year.

60% of respondents say they could contribute between US$ 0.5 and US$ 5 and 15% between US$ 6 and US$ 15; it is expected that 10% of respondents could contribute more than US$ 15 per month.

Figure 3.15 Willingness to pay the additional value

The acceptance of additional value in the bill in exchange for better service and purchase of land from watersheds: 45 consumers, representing 3.80%, are in favor of this approach, 967 people (81.74%) disagree, 12.76% did not know or did not answer and, for 1.69%, this question does not apply.

3.8% are willing to pay an additional US$ 1.22 on average.

3.2.3   Analysis of results

In the market study of 2005, 15% of respondents who answered that their availability of payment in addition to the monthly bill is null. It was interpreted as a form of complaint or reproach against the management, administration and service system.

In the results of the 2011 market study, the perceptions of users changed noticeably; it is indicated as an example the availability of additional payment to the monthly bill that has 1183 users, only 3.8% of them mention that would be willing to pay on average US$ 1.22 per month.

This resounding change could be due to the following circumstances:

1   The municipal administration in the city worsened its performance compared to the one until 2005.
2   Citizens perceived terrible municipal service at the level of drinking water, sewerage, roads and urban decoration.
3   The technical teams of the different dependencies of the municipality did not appraise the present and future urgent need to strategically differentiate them from those less urgent ones, thus affecting the social, urban and financing aspects, with their respective political costs.
4   There were multiple changes in the administration and in its policies.
5   The media campaign of the current political opposition also weighed heavily, which continuously provoked a change in the image of the shift managers and their collaborators and in the services provided by the institution as a whole.

3.2.4   Value of water as the input of production

Given the diversity of uses that water has, economic valuation can be done under the cost-saving approach (social, environmental, others), change in productivity (agricultural and

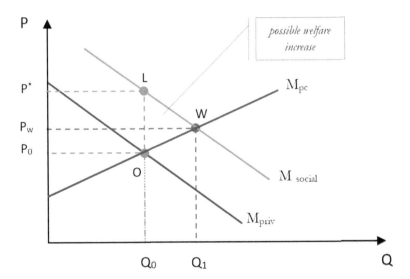

Figure 3.16 Curve of positive externality in consumption

Source: Adapted from http://bit.ly/2Eoug2J; Clark and Lee, 2017

industrial irrigation systems) and consumer surplus (domestic sector). That mixture of valuation approaches provides a differentiated economic value for water when it is used as an input to production.

Externality in consumption

For a positive externality in consumption, Point O (Fig. 3.16) is the equilibrium point for (P_0, Q_0), where the marginal cost of production (M_{pc}) is equal to the marginal private benefit perceived by consumers. From a social point of view, this equilibrium is inefficient since (P^*) the marginal social benefit of consumption Q_0 is higher than P_0.

To optimize the relationship socially, the marginal cost of production must be equal to the marginal social benefit; this occurs with a Q_1 to a P_W.

where
M_{social} The marginal social benefit
M_{priv} The marginal private benefit
M_{pc} The marginal private cost
OLW   Net welfare loss for being in Q_0 and not in Q_1

It follows that the damage caused by the farmers and producers in the lands of the tributary micro-basins – El Carmen and San Simón of the sub-basin of Zamora Huayco – affect Lojana citizenship without the participation of the latter being voluntary and also without direct monetary transactions between them.

From the model of *positive externality in consumption*, the generous and altruistic joy that a person feels when they have reforested and taken care of part of the vegetative components of a micro-basin is proposed as an example, in order to meet the needs of others in the future, such as guaranteeing the quality and quantity of water to future generations, since it generates enthusiasm and satisfaction for those who cared for and reforested early.

By conceptualizing the water resource for production and service, it can be indicated that water would be classified as a standard public natural capital asset, whenever that capital asset is considered to be that produced or co-produced by humankind and it is used for the production of other goods. The particular case of water is included as a source of vital resources, production and essential sustenance for the life and good health of the beings that depend on it.

Notes

1   The market study that was developed in 2004 was based on a sample of 100 users of the potable water service in the entire city of Loja.
2   A part of the 2011 study included the collection of cadastral information for the total number of users residing in three urban neighborhoods in the city of Loja (La Tebaida, Bellavista and San Pedro). This activity involved the surveying of 1183 users of the potable water service in the entire city of Loja.

Chapter 4

Contingent valuation of the water service

Holger Benavides-Muñoz, Jorge Arias-Zari and
José Sánchez-Paladines

4.1 Independent variables (categorical and continuous) vs. the dependent variable

4.1.1 Dependent variable

The availability of incremental payment in the monthly return (aip) is the incremental amount that users – at the date of the analysis – would be willing to pay for the land of the tributary basin to be bought and so that the quality and quantity of water can be improved through the reforestation, protection and maintenance of the area that is the source of water supply (El Carmen and San Simón basins in Zamora Huayco Alto) through an increase in the monthly payment.

4.1.2 Categorical variables

It is known by categorical variables that these reflect the characteristics of the model analyzed; and cannot be written numerically in an integral sense. The values that can be taken is within a small range, thus dividing the characteristics under analysis into categories. These categorical variables in turn are subdivided into:

Nominal: These are the names or characters that do not allow for any numerical interpretation, such as sex, color, race or origin. For its identification an arbitrary number within a range is assigned to each category, as follows: 1 (Ecuadorian), 2 (Brazilian), 3 (Canadian), 4 (Chinese) or 5 (others).

Ordinal: For this type of variable, a numerical (natural) order is attributed to the chosen categories. On the basis of the opinion of the respondents about a new computer product for communication, a variable can be assigned, with values of 1 (bad), 2 (good) or 3 (very good).

Attitudes are not amenable to direct observation but must be assumed or inferred by verbal expressions, by the type of response and by the observed behavior mainly. Depending on the question, values were assigned, with three characteristics estimated between 1 and 3, assessed by attitude scale: a value of 1 for bad, little and/or cheap; a value of 2 for regular, sufficient and/or normal; and a value of 3 for good, very large and/or expensive.

Thus, it is necessary to discern the market information through the correlation between variables, the same ones that lead to the categorical sustenance of the investigation. As an example of application, the variables of the 2004–2005 study are presented next. The relationship of the variable: (gen) gender with (aip), payment availability of everyone surveyed (see Table 4.1). This variable is distributed in dollar ranges: from US$ 0 to US$ 2; from US$ 3 to US$ 5; from US$ 6 to US$ 10; from US$ 11 to US$ 15; and more than US$ 15.

Ratio of the variable: (al) academic level with (aip); a value of 1 is assigned for university person and 0 for others. See Table 4.2.

Relationship of the variable: (age) age with (aip); a value of 1 was assigned for those respondents who are between 17 and 40 years old (since it is the highest labor force rank) and a value of 0 for ages over 40 years. (If they do not have a value, then they do not belong to the economically active population [EAP].) See Table 4.3.

Ratio of the variable: (ardw) attitude rate to the decrease of water with the (aip); a value of 1 was assigned if the amount of water has been reduced and a value of 0 in any other case. See Table 4.4.

Table 4.1 Ratio gen vs. aip (US$)

Variable	Characterization	Availability of incremental payment in the monthly return (aip)					Total
		0–2	3–5	6–10	11–15	> 15	
GEN	Male = 1	21	12	10	1	8	52
	Female = 0	26	16	2	2	2	48
	Total	47	28	12	3	10	100

Table 4.2 Ratio al vs. aip (US$)

Variable	Characterization	Availability of incremental payment in the monthly return (aip)					Total
		0–2	3–5	6–10	11–15	> 15	
Academic Level	University person = 1	34	20	7	3	4	68
	Others = 0	13	8	5	0	6	32
	Total	47	28	12	3	10	100

Table 4.3 Relationship age and aip (US$)

Variable	Characterization	Availability of incremental payment in the monthly return (aip)					Total
		0–2	3–5	6–10	11–15	> 15	
AGE	(17–40) = 1	26	18	9	1	5	59
	(> 40) = 0	21	10	3	2	5	41
	Total	47	28	12	3	10	100

Table 4.4 Ratio ardw vs. aip (US$)

Variable	Characterization	Availability of incremental payment in the monthly return (aip)					Total
		0–2	3–5	6–10	11–15	> 15	
ARDW	(decreased) = 1	45	22	10	3	8	88
	(Other) = 0	2	6	2	0	2	12
	Total	47	28	12	3	10	100

Relationship of the variable: (arvcb) attitude rate to the value currently billed with the (aip), whose signaled characterization is cheap = 1, normal = 2 and expensive = 3. See Table 4.5.

Relationship of the variable: (arbqws) attitude rate before the water service quantity with the (aip), with values of 1 (little), 2 (enough) or 3 (a lot). See Table 4.6.

Relationship of the variable: (arbqlws) attitude rate before the quality of the water service with (aip), where bad = 1, regular = 2 and good = 3. See Table 4.7.

Relationship of the variable: (arbrws) attitude rate before the regularity of the water service with the (aip); thus, a value of 1 is for bad, 2 for regular and 3 for good. See Table 4.8.

Table 4.5 Ratio arvcb vs. aip (US$)

Variable	Characterization	Availability of incremental payment in the monthly return (aip)					Total
		0–2	3–5	6–10	11–15	> 15	
ARVCB	cheap = 1	3	3	1	1	1	9
	normal = 2	28	18	10	2	7	65
	expensive = 3	16	7	1	0	2	26
	Total	47	28	12	3	10	100

Table 4.6 Ratio arbqws vs. aip (US$)

Variable	Characterization	Availability of incremental payment in the monthly return (aip)					Total
		0–2	3–5	6–10	11–15	> 15	
ARBQWS	little = 1	37	21	10	3	7	78
	enough = 2	10	7	2	0	3	22
	a lot = 3	0	0	0	0	0	0
	Total	47	28	12	3	10	100

Table 4.7 Ratio arbqlws vs. aip (US$)

Variable	Characterization	Availability of incremental payment in the monthly return (aip)					Total
		0–2	3–5	6–10	11–15	> 15	
ARBQLWS	bad = 1	36	18	8	2	6	70
	regular = 2	9	9	4	1	4	27
	good = 3	2	1	0	0	0	3
	Total	47	28	12	3	10	100

Table 4.8 Ratio arbrws vs. aip (US$)

Variable	Characterization	Availability of incremental payment in the monthly return (aip)					Total
		0–2	3–5	6–10	11–15	> 15	
ARBRWS	bad = 1	20	10	3	1	4	38
	regular = 2	23	17	7	2	5	54
	good = 3	4	1	2	0	1	8
	Total	47	28	12	3	10	100

Table 4.9 Relation arlpo vs. aip (US$)

Variable	Characterization	Availability of incremental payment in the monthly return (aip)					Total
		0–2	3–5	6–10	11–15	> 15	
ARLPO	do not remove them = 0	3	5	1	0	0	9
	relocate or buy = 1	30	12	7	1	7	57
	expropriate or declare public utility = 2	7	5	3	0	0	15
	other = 3	7	6	1	2	3	19
	Total	47	28	12	3	10	100

Table 4.10 Relation arbmow vs. aip (US$)

Variable	Characterization	Availability of incremental payment in the monthly return (aip)					Total
		0–2	3–5	6–10	11–15	> 15	
ARBMPW	the authorities or administrative = 0	12	3	0	0	4	19
	water quality = 1	11	15	3	2	4	35
	amount of water = 2	13	6	7	1	2	29
	quality + quantity of pipes + networks = 3	10	4	2	0	0	16
	there is no problem = 4	1	0	0	0	0	1
	Total	47	28	12	3	10	100

Relationship of the variable: (arlpo) rate of attitude toward the private owners of the lands of the micro-basins with the (aip); thus, it does not matter or they do not know or do not remove them = 0, relocate or buy = 1, expropriate or declare public utility = 2 and other = 3. See Table 4.9.

Relationship of the variable: (arbmpw) attitude rate before the main problem of the water resource with the (aip). To measure the knowledge about the central issue of the water resource in the city, a value of 0 is assigned in case the problem is related to the authorities or administrative order; 1 if it refers to the water quality; 2 if it is associated with the amount of water; 3 if it is a combination of the quality and quantity of water or networks, pipes, catchment, treatment or distribution; or 4 if there is no problem. See Table 4.10. Relationship of the variable: (arbr) attitude rate before reforestation with (aip). The attitude and knowledge of the reforestation action is valued as 0 if it said that it does not matter or does not know, 1 if reforestation helps water quality, 2 if it increases the quantity and 3 if it attracts the rains and improves the quantity and/or the quality. See Table 4.11.

Relationship of the variable: (occup) occupation level with (aip). The value relates the level of occupation with the availability of payment, where a value of 0 was used for the level of unemployed, student or housework; 1 for a level of artisan, independent or professional technician; and 2 for a level of public or dependent employee. See Table 4.12.

Table 4.11 Relation arbr vs. aip (US$)

Variable	Characterization	Availability of incremental payment in the monthly return (aip)					Total
		0–2	3–5	6–10	11–15	> 15	
ARBR	it does not matter or does not know = 0	4	1	1	1	2	9
	reforestation helps water quality = 1	0	2	6	0	0	8
	it increases the quantity = 2	26	12	0	0	7	45
	it attracts the rains and improves the quantity and/or the quality = 3	17	13	5	2	1	38
	Total	47	28	12	3	10	100

Table 4.12 Relation occup vs. aip (US$)

Variable	Characterization	Availability of incremental payment in the monthly return (aip)					Total
		0–2	3–5	6–10	11–15	> 15	
OCCUP	unemployed, student or housework = 0	10	3	1	0	2	16
	artisan, independent or professional technician = 1	16	15	8	1	6	46
	public or dependent employee = 2	21	10	3	2	2	38
	Total	47	28	12	3	10	100

Table 4.13 Relation ninh vs. aip (US$)

Variable	Characterization	Availability of incremental payment in the monthly return (aip)					Total
		0–2	3–5	6–10	11–15	> 15	
NINH	1 to 3 people	12	7	4	0	1	24
	4 to 6 people	30	20	7	3	7	67
	7 to 10 people	4	1	1	0	1	7
	> 10 people	1	0	0	0	1	2
	Total	47	28	12	3	10	100

4.1.3 Continuous variable

This variable can take an infinity of values, such that for example given three values can take any of such values.

Relationship of the variable: (ninh) number of inhabitants with (aip), which is the number of people living in the respondent's home, grouped by ranges. See Table 4.13.

Relationship of the variable: (mpws) monthly payment for water service sheets with (aip). Six ranges of payment rates are established, between US$ 3 and more than US$ 50. See Table 4.14.

Table 4.14 Ratio mpws vs. aip (US$)

Variable	Characterization	Availability of incremental payment in the monthly return (aip)					Total
		0–2	3–5	6–10	11–15	> 15	
MPWS	(3.0–10.0)	23	10	5	2	0	40
	(10.5–20.0)	14	12	5	1	2	34
	(21.0–30.0)	4	4	2	0	4	14
	(31.0–40.0)	2	1	0	0	2	5
	(41.0–50.0)	2	0	0	0	0	2
	(> 50.0)	2	1	0	0	2	5
	Total	47	28	12	3	10	100

Table 4.15 Relation nha vs. aip (US$)

Variable	Characterization	Availability of incremental payment in the monthly return (aip)					Total
		0–2	3–5	6–10	11–15	> 15	
NHA	(2–4)	7	3	1	0	0	11
	(5–7)	3	0	0	0	0	3
	(8–10)	2	1	0	0	0	3
	(11–13)	1	2	0	0	0	3
	(14–24)	34	22	11	3	10	80
	Total	47	28	12	3	10	100

Table 4.16 Relation mfi vs. aip (US$)

Variable	Characterization	Availability of incremental payment in the monthly return (aip)					Total
		0–2	3–5	6–10	11–15	> 15	
MFI	(< 300)	11	2	1	1	1	16
	(301–500)	18	10	3	0	1	32
	(501–1000)	10	14	4	2	4	34
	(1001–1500)	5	0	4	0	1	10
	(> 1500)	3	2	0	0	3	8
	Total	47	28	12	3	10	100

Relationship of the variable: (nha) number of hours per day of the available water service with (aip). This indicates the number of hours per day that the respondent has the potable water service at home. Ranges are expressed in hours per day (h/d), as follows: from 2 to 4, from 5 to 7, from 8 to 10, from 11 to 13 and from 14 to 24. See Table 4.15.

Relationship of the variable: (mfi) monthly family income with (aip). This measures the amount of money generated in a month, organized into five ranges: less than US$ 300, from US$ 301 to US$ 500, from US$ 501 to US$ 1000, from US$ 1001 to US$ 1500 and more than US$ 1500. See Table 4.16.

Relationship of the variable: (mfe) monthly family expense with (aip). This determines the amount of money spent monthly, organized into five ranges: less than US$ 300, from US$ 301 to US$ 500, from US$ 501 to US$ 1000, from US$1001 to US$ 1500 and more than US$ 1500. See Table 4.17. See Table 4.18 for the distribution of variables according to age.

Table 4.17 Relation mfe vs. aip (US$)

Variable	Characterization	Availability of incremental payment in the monthly return (aip)					Total
		0–2	3–5	6–10	11–15	> 15	
MFE	(< 300)	21	10	4	1	1	37
	(301–500)	14	7	2	0	4	27
	(501–1000)	9	10	6	2	4	31
	(1001–1500)	0	1	0	0	1	2
	(> 1500)	3	0	0	0	0	3
	Total	47	28	12	3	10	100

Table 4.18 Distribution of variables according to age

Variables	Demanders between 17 and 40 years	Demanders > 40 years	Total
Male gender	36.00	19.00	55.00
Availability of incremental payment – average	5.50	6.21	5.86
Academic level – university	47.00	24.00	71.00
Attitude to the decrease in water – perceived to decrease	52.00	36.00	88.00
Attitude toward the value currently billed – normal payment	42.00	23.00	65.00
Attitude toward the quantity of the water service – perceives that it is little	47.00	31.00	78.00
Attitude toward the quality of the water service – perceives that it is bad	43.00	27.00	70.00
Attitude to the regularity of the water service – perceives that it is regular	33.00	21.00	54.00
Attitude toward the private owners of the lands of the micro-watersheds – perceives that the current owners must be bought out or relocated or that the land must be declared as public property or expropriated or similar	51.00	40.00	91.00
Attitude to the main problem of the water resource – perceives that it is the quantity, quality, treatment and/or networks	49.00	31.00	80.00
Attitude toward reforestation – perceives that reforestation increases the quantity and improves the quality of the water or attracts rain	53.00	36.00	89.00
Occupation level – dependent or public employee	32.00	15.00	47.00
Number of inhabitants – a mathematical average	4.30	4.90	4.60
The monthly payment for water service sheets – a mathematical average	18.93	20.59	19.76
Number of hours per day of water service available – water available all day	51.00	29.00	80.00
Monthly family income – a mathematical average	734.39	860.98	797.69
Family monthly expenditure – a mathematical average	477.14	516.83	496.99

Statistical evaluation of pairs of groups among themselves

For the case under study, the research hypothesis can be taken: to determine that the group of claimants between 17 and 40 years of age differs significantly from the group of claimants over 40 years of age. The null hypothesis is that the groups studied do not differ significantly from each other.

Test with contingency tables (χ^2) [chi square]

This method allows for determining whether two categorical variables are related to each other. For this analysis, Equations 4.1 and 4.2 are applied:

$$\chi^2 = \sum \frac{(f_0 - f_e)^2}{f_e} \qquad \text{[Eq. 4.1]}$$

$$gl = (n_f - 1)(n_c - 1) \qquad \text{[Eq. 4.2]}$$

$$f_e = \frac{(row\,total)\,(column\,total)}{sample\,size} \qquad \text{[Eq. 4.3]}$$

where
f_o Frequency observed
f_e Expected frequency
n_f Number of rows
n_c Number of columns

Through the contingency table, Table 4.19, the variables of age and gender are analyzed. The variables of age and gender are independent of each other, which in this case was to be expected. See Table 4.19.

The results in Table 4.20 show that the variables of age and academic level are dependent with 95% confidence, or with the possibility of error of 5%. The results in Table 4.21 show that the variables of age and ARBMWRP (attitude rate before the main water resource problem) are independent with 95% confidence, or with the possibility of error of 5%.

The results in Table 4.22 show that the variables of age and ARBMWRP (attitude rate before the private owners of the watersheds) are independent with 95% confidence, or with the possibility of error of 5%. The results in Table 4.23 show that the variables of age and ARBR (attitude rate before reforestation) are independent with 95% confidence, or with the possibility of error of 5%.

4.2   Financial evaluation – environmental

4.2.1   Value of water in the domestic sector

The fee for the drinking water service in the city under study is structured according to the amount consumed and the user's location or area of residence. Table 4.24 and Table 4.25 are presented as examples of rates. According to the data provided by the Marketing Department of the Municipal Drinking Water and Sewage Unit of the city under study (UMAPAL, 2005;

Table 4.19 Analysis of contingency between age and gender

Gender	VARIABLES	Age		TOTAL
		Users between 17 and 40 years	Users older than 40 years	
	Gender Male	**30.68** 33	**21.32** 19	52
	Gender Female	**28.32** 26	**19.68** 22	48
	TOTALS	59	41	100

CALCULATION OF CHI-SQUARE

f_o	f_e	$(f_o - f_e)^2$	$(f_o - f_e)^2 / f_e$
33	30.7	5.4	0.1754
19	21.3	5.4	0.2525
26	28.3	5.4	0.1901
22	19.7	5.4	0.2735
		$\chi^2 =$	0.8914

INTERPRETATION

gl	Significance	chi^2 critical	Null hypothesis is accepted
1	0.05	3.841	**NOT**
		table E8 (Hanke)	

gl	Significance	chi^2 critical	Null hypothesis is accepted
1	0.01	6.635	**NOT to 99%**

therefore Age and Gender are dependent to 95%

Table 4.20 Contingency analysis between age and academic level

AL VARIABLES	Age		TOTAL
	Users between 17 and 40 years	Users older than 40 years	
AL University person	47 **41.89**	24 **29.11**	71
AL Other	12 **17.11**	17 **11.89**	29
TOTALS	59	41	100

CALCULATION OF CHI-SQUARE

f_o	f_e	$(f_o - f_e)^2$	$(f_o - f_e)^2 / f_e$
47	41.9	26.1	0.6233
24	29.1	26.1	0.8970
12	17.1	26.1	1.5261
17	11.9	26.1	2.1961

$$\chi^2 = 5.2426$$

INTERPRETATION

gl	Significance	chi² critical	Null hypothesis is accepted
1	0.05	3.841	**yes to 95%**

table E8 (Hanke)

gl	Significance	chi² critical	Null hypothesis is accepted
1	0.01	6.635	**NOT to 99%**

therefore Age and AL are dependent to 95%

Table 4.21 Contingency analysis between age and TAPPA

ARBMPW

VARIABLES	Age — Users between 17 and 40 years	Age — Users older than 40 years	TOTAL
ARBMPW Quality + Quantity + pipes + networks	**47.2** 49	**32.8** 31	80
ARBMPW Administrative + There is no problem	**11.8** 10	**8.2** 10	20
TOTALS	59	41	100

CALCULATION OF CHI-SQUARE

f_o	f_e	$(f_o - f_e)^2$	$(f_o - f_e)^2 / f_e$
49	47.2	3.2	0.0686
31	32.8	3.2	0.0988
10	11.8	3.2	0.2746
10	8.2	3.2	0.3951
		$\chi^2 =$	0.8371

INTERPRETATION

gl	Significance	chi² critical	
gl	*Significance*	*chi² critical*	Null hypothesis is accepted
1	0.05	3.841	**NOT**
		table E8 (Hanke)	
gl	*Significance*	*chi² critical*	Null hypothesis is accepted
1	0.01	6.635	**NOT to 99%**

therefore Age and ARBMPW are dependent to 95%

Table 4.22 Contingency analysis between age and ARBMWRP

VARIABLES	Age		TOTAL
	Users between 17 and 40 years	Users older than 40 years	
ARBMWRP To Relocate + to buy + to expropriate	**53.69** 51	**37.31** 40	91
ARBMWRP not to relocate + not to buy	**5.31** 8	**3.69** 1	9
TOTALS	59	41	100

(ARBMWRP)

CALCULATION OF CHI-SQUARE · INTERPRETATION

f_o	f_e	$(f_o - f_e)^2$	$(f_o - f_e)^2 / f_e$	gl	Significance	chi² critical	
51	53.7	7.2	0.1348	1	0.05	3.841	Null hypothesis is accepted **NOT**
40	37.3	7.2	0.1939				table E8 (Hanke)
8	5.3	7.2	1.3627	1	0.01	6.635	Null hypothesis is accepted **NOT to 99%**
1	3.7	7.2	1.9610				

$$\chi^2 = 3.6525$$

therefore Age and ARBMWRP are dependent to 95%

Table 4.23 Analysis of contingency between age and ARBR

	Age		TOTAL
VARIABLES	Users between 17 and 40 years	Users older than 40 years	
ARBR Increases the quality and quantity + attracts the rains	**52.51** 53	**36.49** 36	89
ARBR Does not answer + Does not reforest + other	**6.49** 6	**4.51** 5	11
TOTALS	59	41	100

(Left vertical label: ARBR)

CALCULATION OF CHI-SQUARE

f_o	f_e	$(f_o - f_e)^2$	$(f_o - f_e)^2/f_e$
53	52.5	0.2	0.0046
36	36.5	0.2	0.0066
6	6.5	0.2	0.0370
5	4.5	0.2	0.0532
		$\chi^2 =$	0.1014

INTERPRETATION

gl	Significance	chi² critical	Null hypothesis
1	0.05	3.841	is accepted **NOT**

table E8 (Hanke)

gl	Significance	chi² critical	Null hypothesis
1	0.01	6.635	is accepted **NOT to 99%**

therefore Age and ARBR are dependent to 95%

Table 4.24 Price example for the water service

(US$/m³/month, by category)

Consumption (m³/month)	Category R ($/m³/month)	Category CI ($/m³/month)	Category IN ($/m³/month)	Category O ($/m³/month)	Category OM ($/m³/month)	Category TE ($/m³/month)
10	0.981	0.212	0.315	0.213	0.106	0.049
20	2.667	0.212	0.315	0.213	0.111	0.107
50	7.508	0.335	0.504	0.220	0.128	0.107
70	16.193	0.418	0.629	0.257	0.144	0.132
90	25.610	0.418	0.629	0.257	0.144	0.153
100	36.544	0.418	0.629	0.257	0.144	0.212
200	69.289	0.629	0.946	0.289	0.144	0.253

Note: The categories are R for residential, CI for commercial, IN for industrial, O for official, OM for official average and TE for third age (users who have reached 65 years of age)

Table 4.25 Characteristics of endowments and fees for water service

Category	Average endowment (L/capita/day)	Average price (US$/m³)	Effective micrometering %
Residential	135	0.21	97.43
Seniors	168	0.15	97.80

Source: UMAPAL (2005)

UMAPAL, 2011), the average rates and the provisions typify the more than 30,000 users, for two years later are shown in Table 4.25.

According to this same source, the average percentage of unaccounted-for water of the entire city is 44% (ANC = 44%).

The general average value of water can be obtained by using Equation 4.4.

$$Vp = \frac{\sum_{i=1}^{n} P_i Q_i}{\sum_{i=1}^{n} Q_i}$$ [Eq. 4.4]

where
Vp Total average value of water (US$/m³)
P_i Value of water as an input in category i (US$/m³)
Q_i Water volume demanded in category i (m³/year)

The following is the projection of the general fee for the year 2005 (which includes the consumption and economic collection of all categories):

$$Vp = 0.18176^\$/_{m^3}$$

Thus, the prevailing average rate for the drinking water service in the city under study was 0.18 US$/m³, and for two years later, it was 0.28 US$/m³.

4.2.2   Operation strategy

According to the "Proposal for environmental policies and strategies" of the Environmental Advisory Commission of the Presidency of the Republic (Ecuador) (PAE, 1995), for the viability of the project, it is recommended that the service company itself manage and administer the project for sustainability, with the following recommendations:

1 **Politically – institutionally**

- Establish a municipal policy of openness and involvement in the social organization to participate in municipal decisions on the sustainable development of micro-watersheds.
- Include the variable of community activities in the educational programs of the canton.
- Set guidelines for environmental education policy at all levels of education that tend to form a predisposition toward and encourage decisive participation in the social organization.
- Include in the institutional technical planning a preventive management for the control of water leaks by means of the Infrastructure Leakage Index (ILI) (García-Espinosa & Benavides-Muñoz, 2019).
- Develop and consummate the technical component of communication and environmental tourism for sustainable management.

2 **Economically**

- Establish a fund that covers the costs of technical assistance and training for the right organization.
- Effective use of energy in water supply systems and management of energy efficiency in distribution networks (Benavides-Muñoz & Sánchez, 2010).
- Show statistically that incremental monthly contributions are managed with transparency and that they serve specifically for the maintenance and improvement of the sustainable exploitation of the water resource.

3 **In the empowerment of the project and gender**

- Encourage direct participation of civil society in the project.
- Establish constant mechanisms of support to the social organization for a permanent exchange of opinions and constructive criticisms, establishing direct mechanisms of participation and community opinion regarding environmental management.
- Make the social organization feel like an integral and main part of the diagnoses before the execution of the project.

4 **Legally**

- Revise the legal framework in which the relationship with the community and the environment of the micro-basins is developed.
- Regulate environmental management procedures on the basis of social participation and collective benefit.
- Settle the prices for the purchase and sale of land by using expert opinions, who in turn will make the proposals on the basis of the current cadastral records and a referential basis.
- Ensure that the municipality of Loja uses the type of "exchange" of land or property of its property in the urban area.
- Recommend, under the declaration of "public good," that the micro-watersheds belong in their entirety to the municipality of the city, with the just and timely economic compensation to the current owners.

- Ensure that UMAPAL is responsible for the entire legal process of declaring this whole sector as protective forest of the high slopes of Zamora Huayco.

5   In education

- Include training alternatives in the social organization to find common mechanisms that allow for improvement and access to planning, administration, financing and training methods in environmental management.
- Train leaders to guarantee the continuity and sustainability of the project.

6   Involvement of stakeholders, universities and NGOs

- Have university education institutions and NGOs play a fundamental role since they are research and knowledge transfer entities from and to the public, through a permanent and disinterested study of micro-watersheds.

7   Technically – forestry[1]

- Set up the sowing system so that it can be "houndstooth" or "staggered," with an average distance of 3 m × 3 m. The square is more common, because this system is easy to execute, but the triangular system can also be used; this method has the advantage of 15% more density of sowing.
- Consider the species that serve in reforestation:

  - The bushes like *Carraria thimifolia*; *Bacharis, barbaria sp.*
  - Tasks of stimulation and environmental compensation in humid areas and sandy-silty soils for natural regeneration of the Alder (*Almus jorulensis*).

- Reforest the areas that are currently paddocks and agro-productive farms with native species, such as the Laurel (*Miryca pubscens*); Romerillo (*Podocarpues sprucei*); Piquil (*Gynxys buxifolia*); Quishuar (*Buddleia incana*); Pumamaqui (*Oreopanax sp.*); Chachacomo (*Escallonia mvrtilloide*); Cedro de la Sierra (*Cedrela montana*); Arrayán (*Eugenia sp.*); Quinoa (*Polvlepis incana*); among others.
- Allow the forestry professional to determine the methodology of sowing in each case; here we allow ourselves to briefly recommend a possible way of planting seedlings of tree species. After rubbing, marking and staking in the ground, holes of 45 cm × 45 cm × 40 cm will be excavated. At the time of sowing, it is placed in the pit, surface soil, and between 80 g and 120 g of fertilizer (N, P, K) 12–24–12; or, 80 g to 90 g of 10–30–10, can be added. In addition, in a moderate amount of lime per hole, the seedling of the tree will be placed, with the proper care to plant it at the same depth as that found in the bag or pylon; finally it is watered.
- Weed, prune and exercise foliar control. These tasks will be carried out regularly by the maintenance staff of each micro-basin. It will be developed manually because with machinery the environment is negatively impacted, and we will choose not to use herbicides nor agricultural chemical contaminants. As the primary task of weeding, it is recommended to form the crown; it is necessary to apply manual weeding with a minor tool.
- The protective forest must have visible boundaries and be protected by a fence, preferably natural and alive, with pole-type trees at least two m high, located every 2 m, with five rows of barbed wire, as appropriate.

Additionally, the municipality and the company providing the potable water service can create the position of a park ranger for the surveillance of the micro-basins.

Note

1 This is a summary of the personal dialogue and interviews of the involved technicians and ex-municipal workers of the city under study and of the UMAPAL company.

Chapter 5

Background and environmental problems

Raquel Hernández-Ocampo

Advances in scientific and technological knowledge have undoubtedly helped to create a different consciousness than humanity had before the first half of the twentieth century: human beings have no differences in our respective genomes and there is no racial distinction, and therefore, we are part of humanity and life on the same planet that belongs to all of us, unique in its characteristics to maintain life as we know it. Knowledge supports the voices that arise to recognize us as equal, earthly, culturally diverse, part of nature and therefore part of the multiplicity of trophic chains.

From another perspective, it is necessary to understand that to take advantage of nature in favor of human life, human work is vital to transform nature. Only in the process of photosynthesis can vegetables help to move from inorganic to organic and usable; in this process in which humans intervenes with their intelligence, their workforce is known as the humanization of nature. In this sense, the idea of modifying nature conservatively so as not to affect the requirements of future generations arises.

This long-term view is presented in the report titled "Nuestro Futuro Común. Un resumen" of the Friedrich Ebert Foundation (FFE, 1989), which in the introduction claims,

> During the course of this century has undergone a profound change the relationship sustained between humans and the planet. When the century began neither human beings nor technology had the power to alter the systems of the planet radically. Almost at the end of the century, not only the increase of human beings and their activities have that power, but important unintentional changes are taking place in the atmosphere, the soils, the waters, between plants and the animals and in the relationships that exist between them. The speed with which the change occurs is surpassing the ability of the different scientific disciplines and our capacity to evaluate and advice.
>
> (FFE, 1989, p. 4)

These concerns are placed on political agendas. However, the actions that should be initiated have not started in many parts of the world.

Modernity and modernization are perspectives that establish the directionality of what is now known as globalization, a single dominant model of capitalism, which disseminates the primacy of Western culture, music, dress, the urban design of cities and even a unique food culture, including fast foods, i.e. hamburgers and chicken with potatoes. In the economic area, the profit of the owners of capital interests are paramount, and labor recognition is relegated, such that wages do not cover necessary costs. Under these conditions the possibility of returning to nature what it contributes is far away or that reason, in this system, for a few to achieve success (accumulation without limit), most of the population is reduced to poverty and exclusion.

Industrialism and extractivism (oil and minerals), no matter how much technology and care is used, continue to pollute jungles and seas. Large-scale mining and open sky is an activity that requires large volumes of water that at the end of the process, pollutes flora and fauna, which died, and even severely affects the health of the communities that live in nearby places. Paradoxically, the people who were settled in these places with underground wealth do not benefit from these processes. Rather, the extracted wealth favors large corporations, and the resident population are the so-called liabilities (contaminants) who damage their own health and find themselves poor and excluded, in regional countries such as Perú, Bolivia, and Ecuador, which became open to the intervention of multinational corporations.

In the past few decades, a series of international meetings have been organized by the United Nations to deal with planetary problems such as climate change. Today it is clear that human actions cause the emission of greenhouse gases (GHGs). The excess of individualism in cities requires that each person has a vehicle, and these generate gases that pollute the air. The famous "American dream" leads humanity toward overconsumption and waste, worships things and despises life. The report "Nuestro Futuro Común. Un resumen"

> describes a future that does not work due to current international economic systems, population growth rates, agricultural systems, the increasing number of species in extinction, urban development, the procedures of administration of the "common goods" (the oceans, Antarctica, outer space) and the current weapons culture.
>
> (FFE, 1989, p. 5)

Nature embraces a great diversity of life forms, a few known and most still unknown. By human action, they are irreversibly damaged, because their contribution is unknown and therefore only extracted and not replenished. The disappearance of other species will undoubtedly be the death of the human species. According to FFE (1989, p. 27), and similarly Hinrichsen (1987, p. 30), in the Plain Facts of the Species and Ecosystems: Resources for Development sub-theme, it is indicated that "Estimates of the total number of species that inhabit the earth vary between 5 and 30 million. To date, however, only 1.7 million species have actually been studied and identified." This shows that knowledge is incipient; therefore, it is necessary to discover and study most of the genetic bank that maintains natural systems in its bosom.

Facing this global problem, FFE (1989, p. 7), and Hinrichsen (1987, p. 7) propose that "Together, we must stop using up the earth's ecological capital and start producing on the interest that can be obtained from the sustained management of its resources." To achieve this, it proposes that

> Governments must centralize the concept of sustained development and the community of nations must evolve into a new, fairer international economic structure that begins to close the gap between developed and developing countries. ... This gap in power and resources is the planet's main environmental problem.
>
> (FFE, 1989, p. 7; Hinrichsen, 1987, p. 7).

The report promotes sustained development, understood as "one that meets the needs of the present, without compromising the ability of future generations to meet their own needs" (FFE, 1989, p. 7; Hinrichsen, 1987, p. 7). Regarding needs, it must be understood that they are the basic ones for humanity – food, clothing, housing – complemented by the fundamental one, namely the conservation of nature.

The reading of the complexity of the problems that affect humanity leads us to recognize that hunger affects about one billion people, who do not eat enough to live a healthy life with the ability to work and produce; the report states that "Two thirds of malnourished people live in South-West Asia and one third in the lower African Sahara" (FFE, 1989, p. 13), but on all the continents, despite producing enough food, there is a significant population that bears the social disease of hunger. Agriculture is in crisis, primarily due to the degradation of natural resources, soil erosion on all continents, water and soil contamination due to the excessive use of agro-toxins, deforestation and desertification in Asia, Africa and Latin America.

In the book *Poverty, Desertification and Degradation of Natural Resources*, it is pointed out that "land degradation is a consequence of human action and climatic factors. In the group of human factors, deforestation, excessive extraction of forest products, forest fires, animal overload, over-intensive land use, inadequate management and, finally, the use of technologies not suitable for fragile ecosystems stand out" (Morales, 2005, p. 25), which implies less access to natural resources, thereby deepening the conditions of poverty in the medium and long term.

The concept of poverty proposed by Max Neef exceeds the scope of a need – for example, the misuse of soil and water in the production process is expressed in the level of subsistence (quantity and quality of food), affecting the identity, food sovereignty and the freedom of people belonging to a community. The ways of living, of getting sick and even of dying in a population are an expression of the conditions of life and opportunities or the denial of satisfiers of those needs in the prevailing development model; Neef says that human needs are existential and axiological; among the most important human needs are those of being, having and doing. These in their interrelation and/or articulation define the quality of life.

Humanity supports the shortage of fresh water, due to the modification of the hydrological cycle and especially to the increasing levels of pollution (Arteaga *et al.*, 2020). A considerable number of studies of international organizations affirm that water, in the coming years in many regions of the world, due to its scarcity, will be one of the reasons for conflict between peoples and countries. This situation demands protection and responsible use in the present as part of the planning and integrated management of water resources that considers all types of water sources – quantity, distribution over time and water quality – as well as the actors (public and private) involved and interested in the management of this vital resource for life. Michel Bosquet, quoted in the prologue of the book *Santiago del Estero*, states that "Humanity needed thirty centuries to gain momentum; thirty years are left to stop before the abyss" (Giannuzzo *et al.*, 2005, p. 15), words that summarize the long relationship between humankind and nature and the extreme situation we are facing.

5.1 Environmental problems in the southern region of Ecuador

In the southern region of Ecuador, the degradation of renewable natural resources (soil, vegetation and water) is constant. The estimated deforestation for Ecuador during 2000–2010 was 230,000 hectares/year (ha a^{-1}) (FAO, 2010b, in Ochoa-Cueva *et al.*, 2015). The average annual area deforested between 2016 and 2018 was 82,529 ha, but 24,100 ha a^{-1} also came to regeneration in the same period, so that net deforestation reached 58,429 ha (El Universo, 2019), but it is still the cause that gives rise to other environmental problems, such as the loss of habitats and species. The lack of vegetation is the cause for soil erosion. These two problems, in their mutual impacts, are the two leading causes of the alteration of the quantity and quality of water and of course of the loss of wild organisms. Regarding the protection and maintenance of biological diversity and natural and cultural resources in Ecuador, the first conservation actions in Ecuador date back to 1936, when the country declared the Galápagos Archipelago as a protected area. Since then, conservation policies and strategies have been gaining importance.

As of 1976, the Ministry of Agriculture, through the National Forestry Program and with the support of international cooperation, raised the Preliminary Strategy for the Conservation of Outstanding Wilderness Areas of Ecuador. This strategy marked the beginning of a series of processes and actions tending to consolidate the National System of Protected Areas of Ecuador (SNAP). The procedure laid out the guidelines for moving forward from a predominantly economic governmental vision of "forest resources" toward an approach of protection and conservation of biodiversity, and in 1981, it established the basis for the enactment of the Forest and Conservation Law of Natural Areas and Wildlife.

The Political Constitution of Ecuador promulgated in 1998 gave way to the institutionalization of SNAP in the country, by declaring "the establishment of a National System of Protected Natural Areas that guarantees the conservation of biodiversity and the maintenance of ecological services, in accordance with international conventions and treaties" (CPRE, 1998, p. 25, Art. 86, numeral 3) and specifying the sovereign right of the Ecuadorian state over biological diversity, natural reserves, protected areas and national parks (Article 248). Currently, SNAP is made up of 50 protected areas that occupy a field of 30% of the national territory.

Until 1999, the Ministry of Environment of Ecuador drafted the Strategic Plan of the System of Protected Natural Areas (SNAP), which, although it was not approved, has been a reference for the management of SNAP in the first years of the 2010s. This plan made SNAP visible as a system composed of several subsystems, among them the National Heritage of Protected Natural Areas (PANE), plus the possible and potential areas that will be established by the sectional governments (provincial, municipal, parochial), regional development corporations, communities and the private sector (MAE, 2007, p. 16).

The trend of urbanization determines an imbalance between the urban and the rural populations. In the province of Loja, with the data from the population censuses, the changes that occurred in the period 1950–2010 are reviewed, and the community shows the trend of urbanization: 55% of the population lives in the cities and only 45% in the rural sector (Hernández, 2015). The rural–urban relationship has changed irreversibly, and for this reason, the country population must not only produce to feed those who live in the city but must also conserve water to supply human consumption. Now local governments need to increase urban services and infrastructure, especially for the poor, providing them with the land, affordable building materials, sanitary facilities and drinking water and access to health services, school and public transportation.

Undoubtedly, there are exciting efforts promoted by development projects. These have ventured with participatory technical and methodological alternatives for the management of natural resources, especially in the Andes (the principal aquifer of Amazonia), generating essential proposals that are currently executed. In this context, the scheme of Peasant

Forestry Development, that promoted agroforestry was framed, i.e. the production of short-cycle crops and simultaneously the introduction of the mostly multipurpose tree. This management of the immediate and mediate time of this production process is aimed at achieving economic and social sustainability but also environmental sustainability, especially when planting trees, and given the context that solar energy is used in Ecuador throughout the year.

The project Community Management of the Dry Forests and Micro-basins of the Southwest of Loja, with the support of the government of the Netherlands and under the responsibility of the Dutch Service of Development Cooperation (SNV) and the Ecuadorian Forestry Institute of Natural Areas and Wildlife (INEFAN), carried out several actions between 1997 and 2002. The objective was to contribute to the fight against desertification and the improvement of the living conditions of rural people in the southwestern border area of the province of Loja, through the protection and sustainable management of natural resources. This intervention contributed to the strengthening of the institutional structures and management capacity of municipalities and nongovernmental organizations (NGOs), to promote agroforestry, natural regeneration, woodland pasture (silvopastoril) systems and initiatives that generate benefits for women based on the use of forest products (RAFE, 1998).

In the Chinchipe canton (southeast of Ecuador) and the Jaén province (northern Perú), another project was carried out that provided valuable contributions. It was known as Bosques del Chinchipe and was executed by several entities belonging to Ecuador and Perú. This binational effort recognized and strengthened productive processes such as the cultivation of coffee, which requires association with trees that provide shade (Chamba et al., 2016). Undoubtedly several lessons were learned by the population involved, with which various types of knowledge regarding the recognition of plants with curative abilities of multiple pathologies were valued and necessary agreements were reached to promote the proper management of the forests of the binational Chinchipe basin.

The National University of Loja through the CATER and the NGO Solidarieta Italiana Nel Mondo (COSV), with funds given by the European Union, executed the Concerted Management project for the control of desertification and the regeneration of the dry forest in the cantons of Zapotillo and Macará; the same concern was to control the growing process of desertification and seek the regeneration of the dry forest through a set of technological strategies, the consolidation of peasant groups, institutional strengthening and adequate planning for natural resources, their management and their protection.

The interventions ran while financing was in place. The municipalities were motivated to promote actions toward the proper management of natural resources, and they were interested in the conservation of nature. For this reason, in the Zapotillo zone, it has been possible to recognize new areas of forest as a reserve area; one of them is the La Ceiba Natural Reserve, which protects 10,200 hectares of tropical dry forest and a highly threatened ecosystem in southwestern Ecuador, and is the core area of La Ceiba Conservation and Development Area. This area is the habitat of 31 endemic bird species, of which 11 are endangered, 20 species of amphibians and reptiles and 56 species of trees, such as Guayacán and Hualtaco, that have many traditional uses (NCI, 2013).

Regarding the protection of natural resources in watersheds that provide water for human consumption, a pioneering initiative in Ecuador was the strategy of selling environmental services proposed by the project management of natural resources of Pimampiro canton for the maintenance of the quantity and quality of the water; in the case of the province of Loja,

a similar initiative was launched for the valuation of water for human consumption promoted by the municipality of Celica, an institution that has promoted an ordinance for
the protection of micro-watersheds and other priority areas for the conservation of
natural resources in Celica, due to fact that their natural vegetation covers are in critical
condition. This vegetation cover has had a strong deforestation process, and inadequate
soil use, which has caused a decrease in quality and quantity of water for irrigation and
human consumption (Arteaga *et al.*, 2020).

On the other hand, the management of basins in the south region of the country has as
background the administration carried out by the former Ecuadorian Subcommittee PREDE-
SUR that was created in Washington in 1971, through which Ecuador and Perú subscribed to
the Agreement for the Exploitation of the Binational Hydrographic Basins Puyango-Tumbes
and Catamayo-Chira, which in the 1990s was resumed after the signing of peace between
Ecuador and Perú (Carrera de la Torre, 1990). In this regard, it is known by the inhabitants
of the region that PREDESUR disappeared, and the functions that it fulfilled became the
responsibility of several government entities such as the National Irrigation Institute
(INAR) (for a brief time) and the provincial government.

However, the government investment to promote the protection of watersheds is insufficient. Some resources delivered by external cooperation are channeled for this conserva-
tionist purpose. Today, it is pushed from the Ministry of Environment of Ecuador (MAE).
This initiative is identified as socio-forest and consists in the delivery of economic incentives to peasants and Indigenous communities who voluntarily commit themselves to the
conservation and protection of their native forests, páramo or other natural cover. The
delivery of these incentives is contingent on the protection and conservation of their
forests, which means that people receive the incentive once they meet the monitoring conditions that are determined in agreement with the Environment Ministry, for 20 years.

One of the objectives of the socio-forest proposal is to protect the forests, páramos,
natural vegetation cover, and their ecological, economic and cultural values (around 4
million hectares, equivalent to 66% of the unprotected forests of Ecuador), and another
is to conserve the areas of native forests, páramos and other native vegetation of the
country, reducing deforestation rates (to 50%) and gas emissions of associated greenhouse
effect (generating certificates of reduction of GHG emissions due to avoided deforestation)
(PSB, 2008; MAE, 2013).

The results of this project since its creation in September 2008 are that Socio Bosque has
benefited more than 162,000 people, through more than 2200 conservation agreements on
common lands of Indigenous, peasant and Afro-Ecuadorian communities and individual
owners. Thanks to this mechanism, more than 1,200,000 hectares of native forests and
other high-priority ecosystems are protected (PSB, 2008).

The studies and research carried out confirm that in the province of Loja, the degradation
of the watersheds is due to the anthropic activity of the populations, which support a set of
socio-environmental problems. The emigration of young people from the rural sector determines the depopulation of the countryside of the province of Loja, which in turn affects the
lack of labor. Therefore, the producers (resident population) left in the fields must resort to
inadequate practices, such as the felling and burning of arboreal and shrubby vegetation,
which shortens the time required to prepare the soils.

The Regional Environmental Strategic Plan (PEAR, 2005) carried out by the joint work
of various governmental and nongovernmental institutions, which formed the Regional
Environmental Council (CAR), performs an analysis of environmental issues, which

allowed for proposing strategies and lines of action aimed at improving the environmental management of the provinces of Loja and Zamora Chinchipe. Six management areas are defined in the plan: water resources, soil, biodiversity, forestry development, mining and environmental quality, work that has not yet been recovered to carry out tasks aimed at improving environmental quality in the southern provinces of Ecuador.

Another initiative to care for and protect the water resource began in July 2009, with the participation of the municipal governments of Celica, Loja, Macará, Pindal and Puyango and the Nature & Culture International Corporation, which constituted the administrative mercantile trust Regional Water Fund (FORAGUA) for the conservation, protection and recovery of environmental services and biodiversity of fragile and threatened ecosystems of the provinces of Loja, El Oro and Zamora Chinchipe.

FORAGUA is a mixed trust, public and private, administered by the National Financial Corporation and executed by the constituent municipalities with the permanence of 80 years. This mechanism ensures that local and international resources are invested efficiently, gradually improving the quality and quantity of water for the inhabitants of the region and at the same time protecting the immense natural wealth of southern Ecuador.

This fund has been carrying out environmental compensation projects, purchase of land in the water supply micro-basins, especially in the El Carmen micro-basin, monitoring water quality and restoring degraded areas, with contributions of the environmental taxes that generate around US$ 300,000 annually (NIC, 2009; NIC, 2013).

5.2   Demand for water in El Carmen and San Simón micro-basins

In the last decade (2003–2013), the issue of water supply for human consumption has become one of the central concerns of the inhabitants of the Loja city and, of course, the authorities of the Loja canton. The so-called Drinking Water Master Plan shows that the chosen technical alternative was to capture the water in the Tambo Blanco and Los Leones streams, which belongs to the province of Zamora Chinchipe, from which it is piped to the treatment plant located in the Carigán parish, and then distribute it for human consumption. This reality shows that the growth of the city did not consider the management of the micro-basins that are inside of all of the basin of Loja as water supply sources, and today the administrators of the municipality of Loja must frequently overcome a series of problems that show the difficulty of guaranteeing this vital service (EMAAL EP, 2013).

The destruction of the vegetation cover is one of the leading problems that the water supply micro-watersheds have in the city of Loja, especially in the high parts where there are remnants of forests in their vast majority of native species, which are being altered by agricultural and livestock activities, since these lands are mostly privately owned and their owners have little knowledge about the value of the services provided by the forest and the importance of vegetation as a regulator of water resources.

There is a great need to consider water as a resource of frequent use, in which institutions and the population can value the environmental service and quantify their contribution to social welfare and promote animal and plant life, which will allow for the adequate use of water. It is necessary to achieve a harmonious relationship between humankind and nature, between economic and social development and the defense of natural resources, guaranteeing the provision of water to meet the requirements of future generations.

The water market is one of the frontiers in which even the private ones that manage the market have not ventured, where more than 95% of the water supply services for human consumption are the responsibility of municipalities; however, the lack of safety in its quality influences the growth of the demand for bottled water, which explains the proliferation of operators who make a lucrative business selling water in which the work incorporated to improve quality is incipient (Geo-Loja, 2007).

That is why the municipal GAD of Loja must urgently generate methodological guides that allow for inventorying the water resources (quantity and quality) in all the towns of the Loja canton. The valuation and proposal of using the existing natural resources in these areas, especially of the water resources, would help bring about activities oriented toward the preservation, rational use and restoration of the micro-watersheds.

The Zamora Huayco sub-basin consists of four micro-basins: San Simón, El Carmen, Mendieta and Minas; they constitute the origin of the Zamora Huayco river. This joins at the Bolívar bridge to the north of the city of Loja with the Malacatos river and forms the Zamora river, continuing with this name and later pouring its waters into the eastern region, as initially noted.

The Zamora Huayco sub-basin includes part of the buffer zone of the Podocarpus National Park (PNP) in the Loja canton, being one of the most deteriorated sub-basins because of a marked process of deforestation, inadequate use of the soil, harmful agricultural practices, overgrazing and some forest fires.

In the past 50 years, the foothills of the eastern cordillera of the Andes, especially those of southern Ecuador, due to their climate and vegetation, were constituted in zones of population reception; in 1970s, numerous groups of poor migrants were established, who were expelled from the dry areas of the province of Loja, by the scourge of drought or by the application of the Law of Agrarian Reform and Colonization. In these sites, they reproduced the traditional agricultural practices of the Sierra, which are not suited to these fragile environments; such anthropic activities are the cause of the natural space degradation (Hernández, 2015).

The most important land use is livestock, which is practiced extensively by different groups of producers, as evidenced in El Carmen and San Simón micro-basins. Extensive cattle ranching is characterized by the limited management of cattle; this activity promotes the clearing of large areas of forest for the establishment of pastures. In turn, the inadequate control of the grasslands and the low production of biomass causes the extension of the surface, and the carrying capacity is limited 0.4 UBA/Ha. In these soils, fertility and acidity are not appropriately managed, and acidic soils present severe restrictions to the development of pastures, so it would first be necessary to correct the acidosis.

On the other hand, the rapid expansion of weeds on the grasslands generates a considerable problem: the proliferation of shrubs, when ingested by cattle, causes bovine hematuria. Therefore, after a few years of use, the pastures should be abandoned or renewed, because the weeds invade them quickly, especially the Llashipa (*Pteridium aquilinum*). The overgrazing of cattle in unsuitable areas brings about the compaction of soils, the reduction of infiltration and consequently the increase of surface runoff, which affects the loss of soil fertility (Ochoa-Cueva *et al.*, 2015), increases the costs of production and causes floods and low water levels.

The progressive clearing of the forest causes serious problems in the micro-basins; as for the surrounding areas, the increasing fragmentation of the remnants of the forest hinders the genetic flow in the species, resulting in the loss of flora and fauna. Also, the clearing of the forest leads to an intensification of soil erosion, due to lack of vegetation and soil. The most intense rains, called "aguaceros," are frequent in the study area. They cause surface runoff and landslides, mainly in the areas with the steepest slopes. The torrents of water and debris that flood the lower parts of these basins also cause severe problems to the population that inhabits the lower part.

The intervention of the human being in the studied micro-watersheds shows that the few remaining trees are still being knocked down for extracting wood. The construction of roads, the mining activities in the bed of the streams, gathering drinking water and irrigation are some of the causes that have altered the natural forest cover. Through recent rural–urban migrations and the natural growth of the population, the pressure on the forest increases steadily, because a larger area is needed for construction and agricultural livestock production.

The loss of vegetation cover not only affects runoff and the infiltration regulation of rainwater but also accelerates erosive processes, especially those caused by the runoff of water and by winds. Erosion is one of the most wicked problems and is generated mainly by the performance of harmful agricultural practices. Also, erosion is due to factors such as steep slopes, torrential rains and the opening of roads.

Forest fires can arise as a result of the incidence of some natural factor (lightning, high temperatures, frost, lack of rain, the presence of winds and absence of humidity), which dries the vegetation such that a minimal amount of heat produces combustion and starts a fire. However, it is the anthropic actions, those caused by human beings, such as campfires and burning by farmers, that in recent years has caused fires in the dry season between July and November.

In most cases, such fires would not ordinarily have the devastating effects that they have today, because anthropic actions occur in some cases due to lack of knowledge or due to negligence. For example, the weeding of fields with fire to stimulate the regrowth of pastures affects hundreds of hectares, both because the fire is out of control and because of the damage that these practices cause to the microfauna and the organic matter in the soil.

5.3   Perspectives on this natural environment

The book *Santiago del Estero* in the prologue correctly states that "Accepting environmental problems means recognizing that nature is not an inexhaustible good, but a scarce good, not free and if ever more expensive to protect, not eternal but temporary, which is fragile and runs the risk of disappearing, taking with it, in this extinction, humanity" (Giannuzzo *et al.*, 2015, p. 14). The dominant political systems of the civilized world need to recognize this fragility and discover that human "progress" is not enough to ensure the persistence of life on our planet. "From this conviction, humanity needs to promote conservation actions of nature because in doing so it guarantees its survival" (Giannuzzo *et al.*, 2015, p. 15).

The following ideas are aimed at achieving sustainability in the management of microbasins and the use of natural resources; for this, it becomes necessary to specify what is understood by sustainability. Sustainability has to do with carrying capacity. In line with this aim, we agree that it is better to produce clean crops and use production systems such as agroforestry or silvopastoral, because these prevent the soil from being left without vegetation, thus reducing erosion and also making it possible to take advantage of solar energy, which is available all year.

In the lands that are municipal property, it would be advisable to seek natural regeneration, which is a pertinent suggestion from a social perspective. This can occur in an adjacent area with vegetation similar to the one containing the PNP, where the lands become the buffer of the PNP. Thus, the property owned by the municipality can be converted into tourist attractions and environmental education destinations, with set up trails, shelters and places for bird watching. The design of the facilities could be rustic, with materials used from the natural surroundings to keep harmony between local nature and the desires of tourists and students.

The population living in the sub-basin, duly organized, can become responsible for the care of the banks of the streams, avoiding the entry of vehicles that are washed in the bed of the stream and therefore contaminate the waters with greases and oils. They can also avoid forest fires if they motivate visitors to take certain precautions in the places where they camp.

If it is proposed that the population abandon the agricultural and livestock production activities that use agro-toxins, it is necessary to generate several alternative proposals for entrepreneurship, such as becoming tour guides, offering horse rentals and making handicrafts as souvenirs, for a micro-basin in which nature and the local population live in harmony. It can promote the recovery of species such as Mora, Luma, Quiques and the ancestrally known native plant *Cinchona officinalis*, that has disappeared from these basins.

Chapter 6

Climate change

Andreas Fries

6.1 General overview

As the Fifth Assessment Report (AR5) of the Intergovernmental Panel on Climate Change (IPCC, 2013) indicated, warming of the climate system is definitive, especially since the 1950s, due to increased greenhouse gas (GHG) emissions, which cause increases in atmospheric and ocean surface temperatures. Consequently, global ice and snow cover is decreasing, and the sea level is rising. Regarding the atmosphere, a globally average warming of 0.85°C (0.65°C–1.06°C) was calculated for the period 1880–2012 (Fig. 6.1), on the basis of independent data sets and by applying a linear trend. This trend is confirmed by the analysis of extreme temperatures, which indicate that the number of cold days and nights decrease, whereas the number of warm days and nights increase. The AR5 showed that almost the entire globe has been affected, because the observed temperature increases exceed the natural variability of decadal and interannual fluctuations. Furthermore, changes in precipitation distribution and amounts were found, in which some regions receive more rainfall, whereas other regions receive less rainfall (Fig. 6.2). Besides this, weather has become more extreme, which means that storm intensities and the frequency of long-term droughts increase.

Also, ocean warming increases the energy stored in the climate system and rises the sea level, due to thermal expansion of the water and the additional water input from the melting ice covers. On a global scale, the ocean warming is largest at the upper ocean surface layer (0–75 m), where a temperature increase of 0.11°C (0.09°C–0.13°C) per decade was determined (1971–2010; IPCC, 2013). However, ocean temperature also increased at deeper layers (up to 2000 m), in which 60% of the energy is stored in depths between 0 m and 700 m and 30% in the deeper layer. On a regional scale, changes in the salinity of the sea water were also detected. In regions were evaporation is high, salinity increased, whereas in regions were precipitation amounts increased, water got fresher. This makes evident that evaporation and precipitation distribution over the oceans, and the entire globe, have changed, which may also have influence on the ocean current system (IPCC, 2013).

Regarding ice and snow cover, the Greenland and Antarctic ice shields, as well as most glaciers worldwide, are losing mass, which additionally raises the sea level and changes the global surface albedo. Moreover, mass loss of Greenland and Antarctic ice shields accelerated during the past decade; between 1992 and 2001, an average reduction of 34 Gt yr^{-1} (Greenland) and 30 Gt yr^{-1} (Antarctic) was determined, whereas between 2002 and 2011, the rate of ice loss increased to 215 Gt yr^{-1} and 147 Gt yr^{-1}, respectively. The same is valid for the global glaciers, where the average rate of ice loss increased from

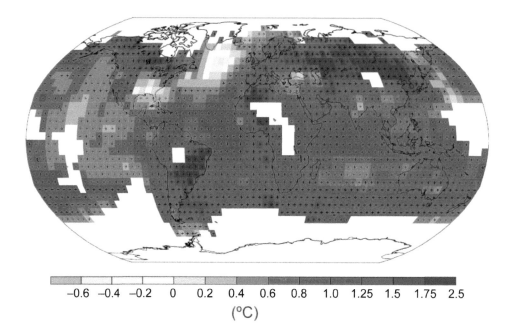

Figure 6.1 Observed surface temperature change from 1901 to 2012, derived from temperature trends determined by linear regression

Source: IPCC, 2013, p. 6

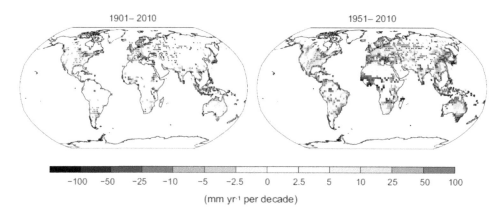

Figure 6.2 Observed precipitation change from 1901 to 2010 and from 1951 to 2010

Source: IPCC, 2013, p. 8

226 Gt yr^{-1} (1971–2009) to 275 Gt yr^{-1} (1993–2009). The ice mass reduction also includes the Artic sea ice extent, which rapidly decreased during past decades, with an average loss of 3.1%–4.1% per decade – in which especially the summer sea ice cover is reduced, by 9.4%–13.6% (Fig. 6.3). Due to the high surface temperature anomalies in the Arctic zone (3°C in Alaska and 2°C in northern Russia), the northern hemisphere snow cover

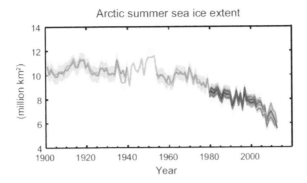

Figure 6.3 Extent of Arctic summer (July–September) average sea ice
Source: IPCC, 2013, p. 10

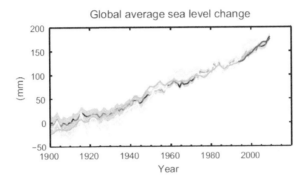

Figure 6.4 Global mean sea level relative to the 1900–1905 mean
Source: IPCC, 2013, p. 10

also decreased, with an average reduction of 1.6% per decade for March and April and 11.7% per decade for June.

Due to the reduction of continental ice and snow cover and the thermal expansion of ocean water, the sea level has risen about 0.19 m over the past century (Fig. 6.4). Additionally, sea level rise has accelerated, as tide-gauge and satellite data indicated. Over the past century (1901–2010), an average increase of 1.7 mm yr^{-1} was calculated, whereas 2.0 mm was estimated for the period 1971–2010, and 3.2 mm yr^{-1} for the period 1993–2010.

The reason for the increase in global warming and its consequences is the emission of GHGs by human activities. The most important GHGs are carbon dioxide (CO_2), methane (CH_4) and nitrous oxide (N_2O), which show concentrations in the atmosphere at levels unprecedented during the past 800,000 years. CO_2 is principally emitted from fossil fuel combustion and cement production (9.5 GtC yr^{-1}) as well as from deforestation and other land use changes (0.9 GtC yr^{-1}), which resulted in a cumulative anthropogenic emission of 555 GtC in 2011 (IPCC, 2013). The cumulative anthropogenic emissions are stored in the atmosphere (240 GtC) and in the oceans (155 GtC) as well as in natural terrestrial ecosystems (160 GtC).

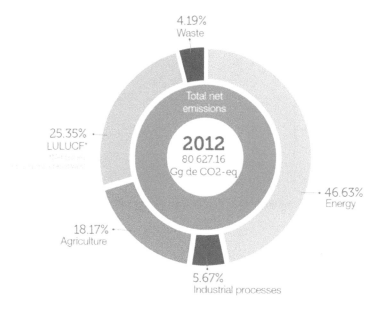

Figure 6.5 Distribution of net emissions in Ecuador in 2012
Source: MAE, 2017, p. 26

The Ecuadorian Ministry of Environment (*Ministerio del Ambiente Ecuatoriano*, MAE) published in 2017 its "*Tercera Comunicación Nacional del Ecuador sobre Cambio Climático*", where the principal sectors of GHG emissions and mitigation actions to face the ongoing global warming are presented on a national scale. Therein, a temperature increase between 0.7°C and 1.0°C since 1970 was determined, and a warming of up to 2°C is expected by 2100.

The report divides the sector of GHG emissions into energy, industrial processes, agriculture, land use change and forestry (LULUCF) and waste, in which all GHG emissions were specified in CO_2-equivalent units ($CO_{2\text{-eq}}$). The total net emission (emission – absorption) of each sector is shown in Figure 6.5, which indicates that the energy sector was the major contributor to national GHG emissions in 2012.

Ecuador is a Non-annex I Party of the United Nations Framework Convention on Climate Change (UNFCCC), and therefore has no mandatory commitments to reduce GHG emissions. Nevertheless, voluntary mitigation actions have been implemented during the period 2011–2015 (MAE, 2017).

In analyzing the total net emission from the different sector during the past decades (1994–2012), only a slight reduction could be stated (Fig. 6.6). However, the proportion of the different sectors had changed in regard to total net GHG emissions, where especially emissions from the energy sector have risen, due to the implementation of new technologies (e.g. smartphones) and population growth (e.g. traffic; Ochoa-Cueva *et al.*, 2015), whereas emissions from the LULUCF sector decreased, due to a reduction in national deforestation rates (MAE, 2012). However, a reduction in national deforestation was not confirmed by other studies, which, in contrast, calculated an increase in deforestation rates in Ecuador (e.g. FAO, 2010a, 2010b; Tapia-Armijos *et al.*, 2015; Gonzalez-Jaramillo *et al.*, 2016).

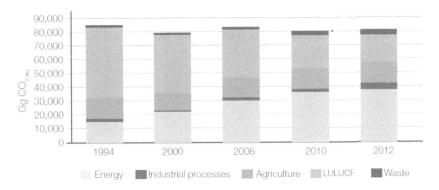

Figure 6.6 Distribution of net emissions in Ecuador in 2012
Source: MAE, 2017, p. 27

6.2 Climate change in Loja

Climate is the average weather conditions at a specific location, which can be determined by a 30-year dataset (climate period; Barry and Chorley, 2010). However, the total of meteorological variables is not available for each station, for which reason generally only temperature and precipitation data are used for climate classifications (e.g. Geiger, 1954). Furthermore, to determine climate change or its respective trends, even longer time series are necessary (more than 30 years). Therefore, to analyze the climate change and its trends in the basin of Loja, the historical information of the station Argelia (1965–2015) were used, specifically the daily temperature (Tmin, Tmean and Tmax) and precipitation (P) data.

First, the daily information was quality controlled, establishing a reliable range for temperature (Tmin, Tmean and Tmax respectively) and precipitation. The temperature range was determined by applying three standard deviations to the mean value (up and down), due to the general normal distribution of temperature data (e.g. Peterson *et al.*, 2001; Zhang *et al.*, 2005). Daily precipitation data were accepted within a range from 0 mm to 70 mm (personal comment, INAMHI Loja), and negative values were eliminated. Afterward, monthly values were calculated if 80% of daily temperature data or 90% of daily P data were available. Finally, annual values were determined if a complete data set of monthly data for a respective year existed.

The annual information was used to generate the overall trends (Tmin, Tmean, Tmax and P), applying a linear regression for the period 1965–2015 (IPCC, 2013). Furthermore, the software R with its extension RClimDex 1.0 (Zhang and Yang, 2004) was used to calculate additional climate indices related to temperature and P on the basis of daily information. The software also provides the respective p-value, which indicates the probability or significance of the calculated trends, which was accepted if p-values were below 0.05.

6.2.1 Temperature trends

All temperature trends (Tmin, Tmean and Tmax) for Loja are positive, indicating a general warming during the past five decades (Fig. 6.7). Tmin increased by $0.025°C\ yr^{-1}$, which results in a total nocturnal temperature increase of 1.25°C. A similar trend was calculated for daily maximum temperatures (Tmax: $0.023°C\ yr^{-1}$), which indicates an increment of 1.15°C over the past 50 years. Average temperatures (Tmean) also increased, but the gradient is weaker (0.012°C yr^{-1}), which is why a general warming of 0.7°C can be stated for Loja.

Besides these temperature trends, by means of the RClimDex 1.0 software, additional indices for Tmin (Fig. 6.8) and Tmax (Fig. 6.9) were calculated on the basis of daily information.

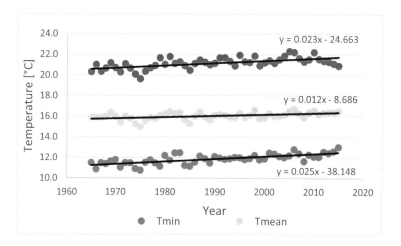

Figure 6.7 Yearly temperature trends over the past 50 years for the city of Loja

(a)

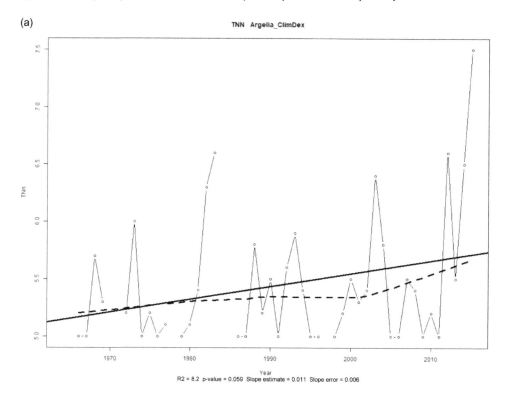

Figure 6.8 Deviated trends of daily Tmin data: (a) TNN, (b) TNX, (c) TN10p and (d) TN90p

(b)

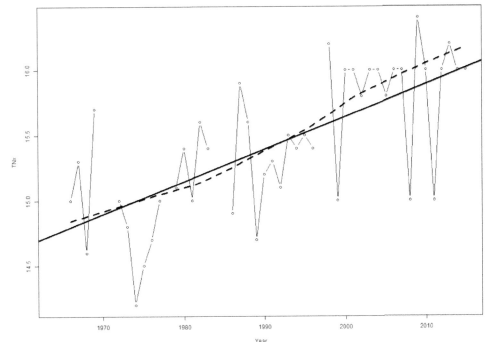

Figure 6.8 (Continued)

(c)

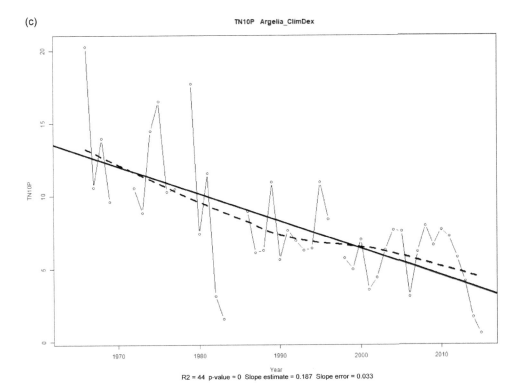

(d)

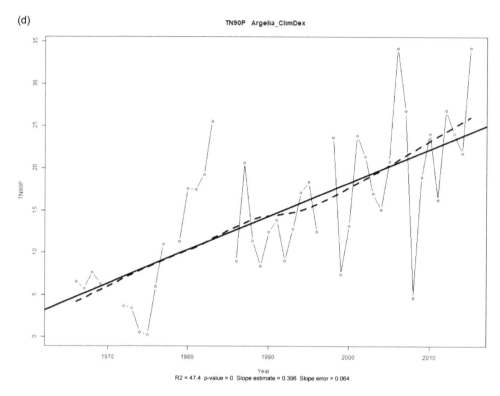

Figure 6.8 (Continued)

The following are the calculated nocturnal Tmin climate indices:

- Nocturnal minimum temperature (TNN [°C]; Fig. 6.8a).
- Nocturnal maximum temperature (TNX [°C]; Fig. 6.8b).
- Proportion of days with minimum temperatures below the 10th percentile (TN10p [%]; Fig. 6.8c).
- Proportion of days with minimum temperature over the 90th percentile (TN90p [%]; Fig. 6.8d).

Similar climate indices also were calculated for daily Tmax by analyzing the highest temperatures during the afternoon:

- Afternoon minimum temperature (TXN [°C]; Fig. 6.9a).
- Afternoon maximum temperature (TXX [°C]; Fig. 6.9b).
- Proportion of days with maximum temperatures below the 10th percentile (TX10p [%]; Fig. 6.9c).
- Proportion of days with maximum temperatures over the 90th percentile (TX90p [%]; Fig. 6.9d).

As shown in Figure 6.8a, the nocturnal minimum temperatures (TNN) increased, which means that nights are not as cold as before. The average trend is $0.011°C \ yr^{-1}$ with a p-value

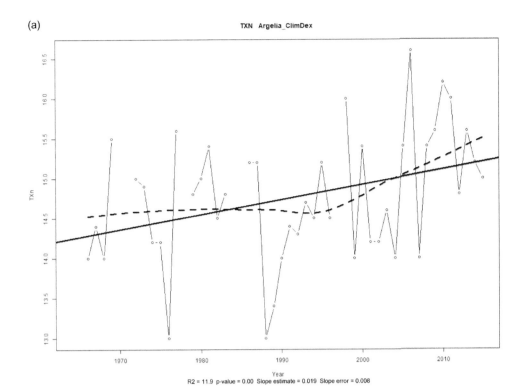

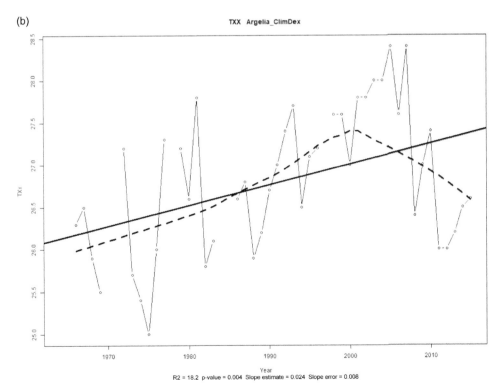

Figure 6.9 Deviated trends of daily Tmax data: (a) TNN, (b) TNX, (c) TN10p and (d) TN90p

(c)

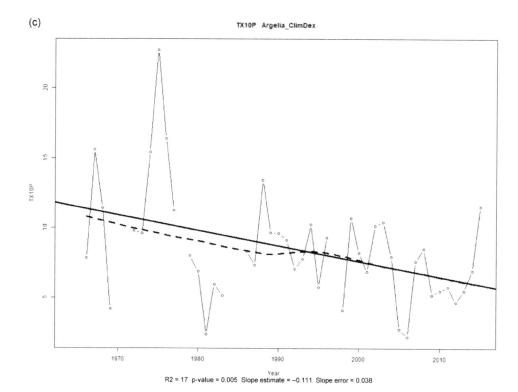

TX10P Argelia_ClimDex

R2 = 17 p-value = 0.005 Slope estimate = −0.111 Slope error = 0.038

(d)

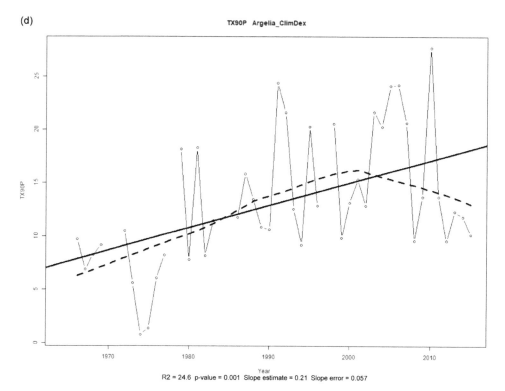

TX90P Argelia_ClimDex

R2 = 24.6 p-value = 0.001 Slope estimate = 0.21 Slope error = 0.057

Figure 6.9 (Continued)

of 0.059, which results in a total increment of 0.55°C for TNN. The calculated p-value is not significant (> 0.05) but close to it, which is why the observed trend can be assumed. Also, the nocturnal maximum temperatures (TNX) increased by 0.025°C yr^{-1} (Fig. 6.8b), which indicates that nights are generally warmer now (1.25°C), which is confirmed by the highest possible significance of p-values (zero). These results are approved by the number of nights with very low temperatures (TN10p; Fig. 6.8c), which show a reduction over the past five decades (–0.187 days yr^{-1}). During the past century, 10–20 nights per year had very low temperatures, but with the beginning of the new century, the number decreased to an average of five nights (p-value = 0). Moreover, the number of nights with high temperatures (TN90p; Fig. 6.8d) clearly increased (0.396 days yr^{-1}; p-value = 0) from an average of 15 days per year at the end of the past century to 20–25 days per year during the past decade.

The trends of daily Tmax (Fig. 6.9), which occur in the afternoon, show the same behavior as Tmin. Daily minimum Tmax (TXN) increased, with an average gradient of 0.019°C yr^{-1} (Fig. 6.9a), which means that the lowest Tmax values in the afternoon are 0.95°C higher than those at the beginning of the observations period, which is confirmed by a significant p-value of 0.022. Also, daily maximum Tmax (TXX) increased significantly (p-value: 0.004), with 0.024°C yr^{-1} (Fig. 6.9b), indicating a total increase of 1.2°C. Similar to Tmin, the number of days with low Tmax decreased (–0.111 days yr^{-1}; p-value: 0.005), from 12 days per year on average during the past century to seven days per year (Fig. 6.9c). The higher temperatures in the afternoon are also confirmed by the number of very hot days (TXX; Fig. 6.9d), which increased significantly, by 0.210 days yr^{-1} (p-value: 0.001), reaching 20 days per year on average.

In summary, nocturnal Tmin in Loja increased by about 1°C; the number of warm nights increased, whereas the number of cold nights decreased significantly. Concurrently, Tmax increased significantly (~1°C) as did the number of very hot days, which confirms the general warming of the basin.

6.2.2 Precipitation trends

Annual precipitation amounts in Loja increased during the observation period (1965–2015; Figs. 6.10a and 6.10b). The calculated general gradient is 3.92 mm yr^{-1} (Fig. 2.25a), which indicates a total increase of 196 mm over the past 50 years. This trend is confirmed by the RClimDex 1.0 software, which calculated a general gradient of 4.02 mm yr^{-1} (Fig. 6.10b). Both charts display an average annual precipitation of about 800 mm at the beginning of the observation period, which increased to more than 1000 mm during the past decade.

To analyze the increases in precipitation amounts for Loja, three additional trends or climate indices were calculated by means of the RClimDex 1.0 software (Fig. 6.11):

- Number of days with precipitation over 10 mm (R10mm [#days]; see Fig. 6.11a).
- Number of days with precipitation over 20 mm (R20mm [#days]; see Fig. 6.11b).
- Number of days with precipitation over the 95th percentile (R95p [#days]; see Fig. 6.11c).

The selected trends show overall positive gradients, indicating that days with precipitation over 10 mm and over 20 mm have increased, as have days with extreme rainfall events

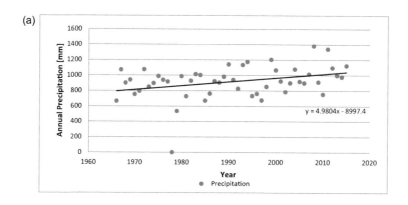

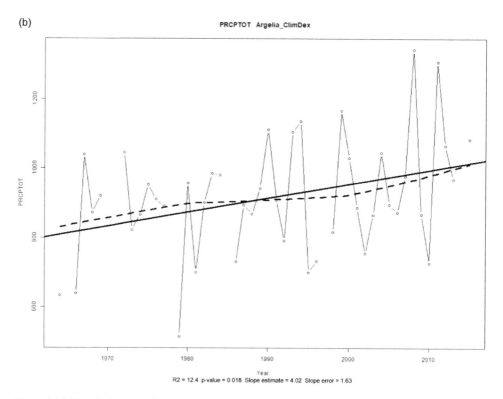

Figure 6.10 Trends for annual precipitation: (a) personal calculation and (b) RClimDex 1.0

(R95p). However, the calculated trends are significant only for R20mm (p-value: 0.005; Fig. 6.11b) and R95p (p-value: 0.011; Fig. 6.11c), which indicates that especially extreme events become more frequent. Therefore, it can be concluded that the increase in P is principally caused by extreme events, which is especially problematic for civil structures and the drinking water supply.

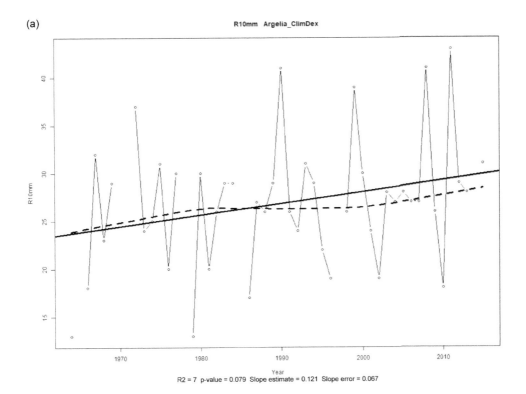

(a)

R10mm Argelia_ClimDex

Year
R2 = 7 p-value = 0.079 Slope estimate = 0.121 Slope error = 0.067

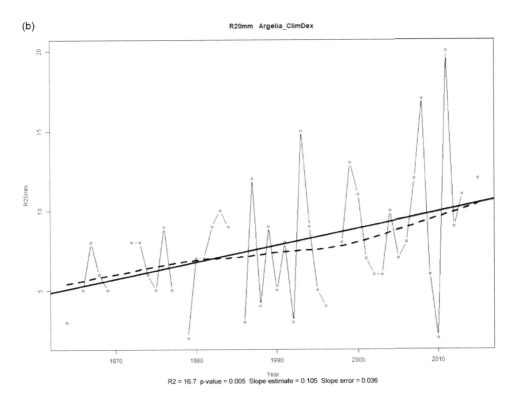

(b)

R20mm Argelia_ClimDex

Year
R2 = 16.7 p-value = 0.005 Slope estimate = 0.105 Slope error = 0.036

Figure 6.11 Deviated trends of daily P data: (a) R10mm, (b) R20mm and (c) R95p

(c)

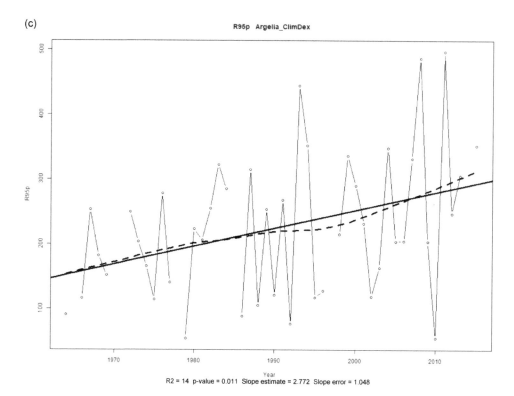

Figure 6.11 (Continued)

6.2.3 Changes in the climate of Loja during the past 50 years

To illustrate the calculated trends and the changes in the climate of Loja, the monthly average temperature (Tmin, Tmax and Tmean) and precipitation (P) data from 1966 to 1995 (Table 6.1) are compared to the actual climate (1986–2015; Table 6.2). Comparing the tables, all T variables clearly increased, especially the extreme values (Tmin and Tmax). Respective to Tmin, all average monthly values increased between 0.4°C and 0.6°C during the observation period. At the end of the past century, monthly Tmin was generally below 12.0°C (Table 6.1), whereas at the beginning of the new century, monthly Tmin were mostly over 12.0°C (Table 6.2). The increase is especially obvious during the transition months (Oct/Nov) between austral winter (dry season) and austral summer (rainy season). This change may be explained by higher cloud frequencies during the nights, impeding nocturnal outgoing radiation and rising nocturnal temperatures, which is confirmed by increased precipitation amounts. In consequence, the average yearly Tmin has increased during the past 50 years (1966–1995: 11.6°C; 1986–2015: 12.1°C).

Monthly Tmax values increased too, but these were different from Tmin, with more pronounced variations between the seasons. Small Tmax increases (between 0.1°C and 0.4°C) were observed for austral summer (Nov–May), because during the rainy season, cloud frequency is generally high, which prevents temperature extremes. In general, cloud cover has negative effects on Tmax, because clouds reflect the incoming solar radiation and reduce

Table 6.1 Average monthly T and P values for the period 1966–1995

Month	Tmin	Tmax	Tmean	P
Jan	11.8	20.8	15.9	97.0
Feb	11.9	21.2	16.1	120.1
Mar	12.0	21.2	16.1	144.9
Apr	12.0	21.5	16.3	91.5
May	11.9	21.1	16.1	53.8
Jun	11.7	20.0	15.5	53.8
Jul	11.5	18.8	14.8	57.8
Aug	11.5	19.5	15.3	46.9
Sep	11.6	20.7	15.8	39.4
Oct	11.1	22.1	16.2	71.3
Nov	11.0	22.7	16.5	60.5
Dec	11.6	22.2	16.4	84.0
Year	**11.6**	**21.0**	**15.9**	**921.1**

Table 6.2 Average monthly T and P values for the period 1986–2015

Month	Tmin	Tmax	Tmean	P
Jan	12.2	21.3	16.2	98.2
Feb	12.4	21.3	16.2	131.9
Mar	12.5	21.6	16.4	147.9
Apr	12.4	21.9	16.5	95.8
May	12.4	21.5	16.3	67.2
Jun	12.2	20.3	15.7	62.3
Jul	12.0	19.6	15.2	52.9
Aug	11.9	20.1	15.6	42.0
Sep	12.0	21.5	16.2	40.9
Oct	11.7	22.8	16.6	76.0
Nov	11.6	23.1	16.6	73.8
Dec	12.0	22.5	16.5	89.5
Year	**12.1**	**21.4**	**16.2**	**978.4**

daily maximum temperatures. Therefore, the increase of Tmax during austral summer is reduced but still detectable. Nonetheless, the increased Tmax during austral summer leads to higher PET, which is why precipitation amounts (P) also increased, as shown in Table 6.1 and Table 6.2. This was indicated previously with the additional rainfall, received mainly by extreme events (R20mm and R95P; Section 6.2.2).

Highest Tmax increases were calculated for austral winter, the relative dry season (Jun–Oct), displaying differences between 0.6°C and 0.8°C. These strong Tmax increases may be explained by reduced P during the extreme months in austral winter (July and August; Table 6.1 and Table 6.2), which indicate lower cloud frequency during this period, for which reason more solar energy is received. The monthly Tmax increment resulted in an average yearly Tmax increase of approximately 0.4°C (1966–1995 was 21.0°C; 1986–2015 was 21.4°C).

The increase of Tmin and Tmax consequently leads to an increase in Tmean (1966–1995 was 15.9°C; 1986–2015 was 16.2°C). However, monthly average warming (Tmean) is lower (0.1°C to 0.4°C) compared to Tmin and Tmax, because for its calculation, all measured daily Tmean data are included (average), not only extreme values (Tmin and Tmax). The highest increase of Tmean was calculated for austral winter (Jul–Oct) because of the strong Tmax rise (up to 0.8°C) caused by the reduced cloud cover and rainfall. Nonetheless, total annual precipitation amounts (P) have increased notably by 57.3 mm, which is due to higher rainfall amounts during austral summer, caused by the increase of extreme events (Nov–Apr). In austral winter (Jun–Sep), rainfall slightly decreased, but a considerable annual precipitation surplus from austral summer remains, which amplifies the differences between the seasons.

Chapter 7

Conclusions and recommendations

Holger Benavides-Muñoz

7.1 Conclusions

The results of the market survey of the year 2004–2005 differ with those of the 2011 market study. These differences are as follows:

1 2004–2005 study: the general average of payroll is US$ 19.60 per month per client (here predominantly the owners of residential houses and centric housing uses, intended for room rental, offices and mini apartments).
2 2011 study: in a general average of payment per spreadsheet of water consumption is of US$ 17.66 per month per client (here predominantly the owners of houses with uses of urban single-family housing, with few houses for rent).
3 From the market study of 2004–2005, it can be deduced that the stratum with an IPR per monthly payroll between US$ 3 and US$ 5 is located within the division of groups comprised as follows: "University students, between 17 and 40 years old, who considers that the current payroll payment is normal, and the amount and quality is small and limited." After the study and after having handled statistically the information collected, the only thing that does not match with what initially was proposed in the hypothesis was cheap payment; instead, the university students considered it a standard amount (neither expensive nor cheap).
4 2004–2005 study: 10% of respondents are willing to pay more than US$ 15; and in turn, of these 10.8% have a monthly income between US$ 301 and US$ 1000.
5 Of the five respondents who exceed US$ 1000 in monthly income, only one is willing to pay more than US$ 15. The three respondents who earn more than US$ 1500 have a DPI between US$ 0 and US$ 2.
6 Of the sample, 91% of the respondents agree that it is purchased or expropriated, or only administered by the municipality through the company providing the service.
7 2004–2005 study: the incremental availability of monthly payroll is US$ 5.86.
8 2011 study: 3.8% of the 1183 respondents would be willing to pay on average US$ 1.22 per month.
9 A limitation to ensure the sustainability of the potable water supply system for the city under study is the quantity and quality of water available now and the imminent need to manage the financial resources for the acquisition of the farms, to maintain and control the micro-basins permanently.
10 The perception that users have of the scarcity of the water resource and its poor quality in the distribution system is directly proportional to the value assigned through the DIP in the market study.

11   The population is convinced that regardless of their occupational condition, monthly income, graduation or academic degree, reforestation increases the quantity and improves the quality of the water; also, they know that it attracts rainfall and that the vegetation cover serves to store and accumulate water in the basin.

7.2   Recommendations

1   Consider a weighted rate according to the strata and willingness to pay incremental group.
2   Maintain strategic agreements with academic, governmental and possibly military institutions to jointly plan and initiate a massive campaign to support the management of the basin, reforest and permanently take care this environment.
3   Promote eco-tourism projects for the sustainable management of the source basins and their water production.
4   Add a follow-up and planned control of the activities to this process. This monitoring of operations and oversight for compliance with indicators will allow us to promptly detect the shortcomings of the project.
5   Develop several activities:

- Legal process for acquiring lands and farms, land valuation, compensation and payment.
- Afforestation, reforestation and natural regeneration.
- The purchase, transport and planting of native trees.
- Maintenance and timely control.
- Enclosures and/or protection in areas of natural regeneration.
- Designs and budgets for gabions, enclosures and signs.
- Protection for slopes and channels.
- Implementation of a training plan for families settled near the area of interest.
- Monitoring and periodic evaluation.
- Dissemination of results and citizen participation.

Bibliography

Andrade, O., Kappas, M. & Erasmi, S. (2010) Assessment of erosion hazard in Torres municipality of Lara State (Venezuela) based on GIS. *Interciencia*, 35, 348–356.

Arteaga, J., Ochoa, P., Fries, A. & Boll, J. (2020) Identification of priority areas for integrated management of semi-arid watersheds in the Ecuadorian Andes. *Journal of the American Water Resources Association* (in press). https://doi.org/10.1111/1752-1688.12837

Ayala-Carcedo, F.J. & Olcina Cantos, J. (Coord.) (2002) *Riesgos naturales*. Editorial Ariel, Ariel Ciencia, 1ª edición, Barcelona. 1512p.

Barry, R.G. & Chorley, R.J. (2010) *Atmosphere, Weather and Climate* (9th ed.). Routledge, Taylor and Francis Group, London and New York. ISBN 13: 978-0-415-46569-4 (hbk)

Benavides, J.D. & Solano, C.C. (2005) *Evaluación del impacto ambiental de los planes de forestación y reforestación ejecutados en la cuenca del Zamora Huayco del cantón Loja, provincia de Loja*. Tesis de Ingeniería en Gestión Ambiental, UTPL, Loja. p. 193.

Benavides-Muñoz, H. (2010) *Diagnóstico de la sostenibilidad de un abastecimiento de agua e identificación de las propuestas que la mejoren*. Tesis doctoral, Universidad Politécnica de Valencia – UPV, España. Disponible de: http://bit.ly/HMBM_TD_UPV

Benavides-Muñoz, H. (2018) *Evaluación y diagnóstico de sistemas de abastecimiento de agua. Sostenibilidad: social, económica y ambiental*. Editorial Académica Española, EAE, España. ISBN-13: 978-620-2-14728-6. 352 p.

Benavides-Muñoz, H. & Arias, J. (2005) *Programa de economía ambiental para la explotación sustentable y sostenible de agua de la cuenca del Zamora Huayco, quebradas El Carmen y San Simón – Loja – Ecuador*. Tesis de Maestría, Escuela Superior Politécnica del Litoral, ESPOL.

Benavides-Muñoz, H. & Paltín, G. (2011) *Catastro y sectorización para el rediseño de la red de distribución de agua potable para el distrito hidrométrico La Tebaida y su propuesta de gestión mediante monitoreo remoto y virtual (ampliación del Laboratorio Virtual de Fluidos, Hidráulica y Eficiencia Energética)*. Proyecto Intranetcittes 2011–2012, Sección de recursos Hídricos, Universidad Técnica Particular de Loja – UTPL, Ecuador.

Benavides-Muñoz, H. & Sánchez. J. (2010) *Gestión para la eficiencia hidráulica y energética en sistemas de distribución de agua*. Libro Sistemas de Saneamento Eficiência Energética. ISBN: 978-85-7745-607-9, cap. 13. Editora Universitaria – UFPB, Joao Pessoa, Brasil.

Bendix, J. & Lauer, W. (1992) Die Niederschlagsjahreszeiten in Ecuador und ihre klimadynamische Interpretation (in German). *Erdkunde*, 46, 118–134.

Bendix, J., Rollenbeck, R. & Palacios, E. (2004) Cloud classification in the tropics: A suitable tool for climate ecological studies in the high mountains of Ecuador. *International Journal of Remote Sensing*, 25(21), 4521–4540.

Bendix, J., Trachte, K., Cermak, J., Rollenbeck, R. & Nauß, T. (2009) Formation of convective clouds at the foothills of the tropical eastern Andes (South Ecuador). *Journal of Applied Meteorology*, 48, 1–17.

Carrera de la Torre, L. (1990) *El proyecto binacional Puyango Tumbes*. Asociación de Funcionarios y Empleados del Servicio Exterior (AFESE-ILDIS), Quito, Ecuador. 246 p.

Chamba, Y., Daniel Capa, E. & Ochoa, P. (2016) Environmental characterization and potential zoning for coffee production in the Andes of southern Ecuador. *Paper presented at the 21st Century Watershed Technology Conference and Workshop Improving Water Quality and the Environment Conference Proceedings*, St. Joseph, MI.

Cifuentes-Carvajal, A. (2016) *Interpolación de lluvia anual para el departamento de Caldas (Colombia)*. Universidad de Manizales, Colombia.

Clark, J.R. & Lee, D.R. (2017) Too inexpensive to be inexpensive: How government censorship increases costs by disguising them. In: *Explorations in Public Sector Economics*. Springer, Cham. pp. 35–50.

ClimaTemps (2018a) *Guayaquil Climate & Temperature*. Available from Homepage: www.guaya quil.climatemps.com/index.php [accessed 30 October 2018].

ClimaTemps (2018b) *Puyo Climate & Temperature*. Available from Homepage: www.puyo.clima temps.com/index.php [accessed 30 October 2018].

ClimaTemps (2018c) *Quito Climate & Temperature*. Available from Homepage: www.quito.clima temps.com/index.php [accessed 30 October 2018].

CPRE (1998) *Constitución Política de la República del Ecuador*. Registro official: 1 de 11 de agosto de 1998, Ecuador.

Du, J., Shu, J., Liu, C., Guo, Y. & Zhang, L. (2012) Variation characteristics of reference crop evapotranspiration and its responses to climate change in upstream areas of Yellow River basin. *Transactions of the Chinese Society of Agricultural Engineering*, 28(12), 92–100.

El Universo (2019) *Casi 60 mil hectáreas al año quedan deforestadas en Ecuador*. Publicado: 24, noviembre, 2019. Editado por Grupo El Universo. Disponible de: https://www.eluniverso.com/noticias/2019/11/24/nota/7616639/deforestacion-ecuador-2019-consecuencias

EMAAL EP (2013) *Informe de actividades de la Empresa Pública Municipal de Alcantarillado y Agua Potable de Loja*, EMAAL-EP. Gobierno Autónomo Descentralizado Municipal de Loja.

FFE (Fundación Friedrich Ebert) (1989) *Nuestro futuro común. Un resumen*. Fundación Friedrich Ebert, World Commission on Environment and Development & Brundland, México. p. 39.

Food and Agriculture Organization of the United Nations (FAO) (2000) *Informe sobre la Disponibilidad de agua en el mundo*. Disponible de: www.fao.org/landwater/aglw/aquastabweb/dbase/html [Accessed July 21, 2002].

Food and Agriculture Organization of the United Nations (FAO) (2010a) *Evaluación de los recursos forestales mundiales 2010*: Informe Nacional Ecuador, Roma. ISBN: 978-92-5-306654-4. Disponible de: www.fao.org/docrep/013/i1757s/i1757s.pdf [accessed 30 October 2018].

Food and Agriculture Organization of the United Nations (FAO) (2010b) *Global Forest Assessment 2010: Main Report*. Food and Agriculture Organization of the United Nations (FAO) Forestry Paper 163, Rome.

Foresti, L. & Pozdnoukhov, A. (2012) Exploration of Alpine orographic precipitation patterns with radar image processing and clustering techniques. *Meteorological Applications*, 19, 407–419.

Fries, A., Rollenbeck, R., Bayer, F., Ganzalez, V., Oñate-Valdivieso, F., Peters, T. & Bendix, J. (2014) Catchment precipitation processes in the San Francisco valley in southern Ecuador: Combined approach using high resolution radar images and in-situ observations. *Meteorology and Atmospheric Physics*, 126(1–2), 13–29. Doi:10.1007/s00703-014-0335-3

Fries, A., Rollenbeck, R., Göttlicher, D., Nauß, T., Homeier, J., Peters, T. & Bendix, J. (2009) Thermal structure of a megadiverse mountain ecosystem in southern Ecuador, and its regionalization. *Erdkunde*, 63(4), 321–335.

Fries, A., Rollenbeck, R., Nauss, T., Peters, T. & Bendix, J. (2012) Near surface air humidity in a megadiverse Andean mountain ecosystem of southern Ecuador and its regionalization. *Agricultural and Forest Meteorology*, 152, 17–30.

Fries, A., Silva, K., Pucha-Cofrep, F., Oñate-Valdivieso, F. & Ochoa-Cueva P. (2020) Water balance and soil moisture deficit of different vegetation units under semiarid conditions in the Andes of southern Ecuador. *Climate*, 8(2), 30. Doi:10.3390/cli8020030.

Foundation Bustamante De La Fuente (2010) Cambio climático en el Perú-Amazonía. *Apus Graph Ediciones*, Lima, Perú.

Fuentes Junco, J.J.A. (2004) *Análisis morfométrico de cuencas: caso de estudio del Parque Nacional Pico de Tancítaro. Dirección General de Investigación de Ordenamiento Ecológico y Conservación de Ecosistemas.* Instituto Nacional de Ecología – INE, México. Publicación especial. 47 p.

Gallegos Reina, A. (2015) *Caracterización de cuencas fluviales periurbanas con riesgo de inundación en ámbitos mediterráneos y propuesta de cartografía de peligrosidad adaptada.* Tesis doctoral, Universidad de Málaga, Disponible en la base de datos TESEO. Disponible de: http://riuma.uma.es/xmlui/handle/10630/10576.

Gallegos Reina, A. (2018) *Caracterización y análisis de los riesgos naturales en el planeamiento urbanístico del litoral mediterráneo español.* Ed. Universidad de Málaga. ISBN: 978-84-17449-18-6

García-Espinosa, J.C. & Benavides-Muñoz, H. (2019). Adjustment value of water leakage index in infrastructure. *Dyna*, 86(208), 316–320.

García García, E. & Ojeda Ochoa, O. (1994) *Estudio sobre la erosión y protección de la micro cuenca de Zamora Huayco, proveedora de agua potable para la ciudad de Loja.* Instituto de Investigaciones Agrícolas. Universidad Nacional de Loja, Ecuador.

Geiger, R. (1954) Klassifikation der Klimate nach W. Köppen (in German). In: *Landolt-Börnstein – Zahlenwerte und Funktionen aus Physik, Chemie, Astronomie, Geophysik und Technik, alte Serie,* Volume 3. Springer, Berlin. pp. 603–607.

Geo-Loja (2007) *Perspectivas del Medio Ambiente Urbano.* Programa de las Naciones Unidas para el Medio Ambiente – PNUMA, NIC, GAD-Loja, Ecuador. p. 191. Disponible de: http://www.natur alezaycultura.org/docs/Geo%20Loja.pdf

Giannuzzo A.N., Ludeña M.E., Contreras J., Cavallotti A. & Barquini L. (2005) *Santiado del Estero: Una mirada ambiental.* Editorial Brujas, UNSE, Córdoba, Argentina. ISBN: 987-99083-9-2

González-Jaramillo, V., Fries, A., Rollenbeck, R., Paladines, J., Oñate-Validvieso, F. & Bendix, J. (2016) Assessment of deforestation during the last decades in Ecuador using NOAA-AVHRR satellite data. *Erdkunde*, 70(3), 217–235. Doi:10.3112/erdkunde.2016.03.02

Gorrab, A., Simonneaux, V., Zhribi, M., Saadi, S., Baghdadi, N., Chabaane, Z.L. & Fanise, P. (2017) Bare soil hydrological balance model "MHYSAN": Calibration and validation using SAR moisture products and continuous thetaprobe network measurements over bare agricultural soils (Tunisia). *Journal of Arid Environments*, 139, 11–25. Doi:10.1016/j.jaridenv.2016.12.005

Hernández, F. (2015) *Situación Agraria y Desarrollo de Loja.* Universidad Nacional de Loja, Ecuador. p. 256. ISBN: 978-9978-355-28-2.

Hinrichsen, D. (1987) *Our Common Future – A Readers Guide. The Brundtland report explained* (2nd ed). Earthscan, Washington, DC. ISBN: 1-85383 0100 (4th printing 1989).

INAMHI (1990–2013) *Anuarios Meteorológicos.* Quito, Ecuador. Available from Homepage: www.serviciometeorologico.gob.ec/biblioteca/ [accessed 30 October 2018].

Instituto Espacial Ecuatoriano (IEE) (2013) *Memoria Técnica Cantón Loja: Geopedología.* Proyecto Generación de geoinformación para la gestión del territorio a nivel nacional escala 1: 25 000. Quito, Ecuador.

IPCC (2013) *Cambio climático 2013, bases físicas. Resumen para responsables de políticas.* Grupo de Trabajo I. Contribución del Grupo de Trabajo I al Quinto Informe de Evaluación del Grupo Intergubernamental de Expertos Sobre el Cambio Climatico (IPCC). Disponible de: http://www.ipcc.ch/pdf/assessment-report/ar5/wg1/WG1AR5_SPM_brochure_es.pdf [accessed December 15, 2014]. ISBN: 978-92-9169-338-2

Johansson, B. & Chen, D. (2005) Estimation of areal precipitation for runoff modelling using wind data: A case study in Sweden. *Climate Research*, 29, 53–61.

Körner, C. & Paulsen, J. (2004) A world-wide study of high altitude treeline temperatures. *Journal of Biogeography*, 31(5), 713–732.

León-Gómez, M., Símuta-Champo, R., Vázquez-Montoya, I. & Solano-Barajas, R. (2016) Análisis comparativo de los métodos para interpolar precipitación en el estado de Chiapas. *Lacandonia*, 10(2), 53–60.

MAE Ministerio del Ambiente del Ecuador (2007) *Plan Estratégico del Sistema Nacional de Áreas Protegidas del Ecuador 2007–2016*. Informe Final de Consultoría. Proyecto GEF: Ecuador Sistema Nacional de Áreas Protegidas (SNAP-GEF), REGAL-ECOLEX, Quito

MAE Ministerio del Ambiente del Ecuador (2012) *Línea base de deforestación del Ecuador Continental*. Proyecto Socio Bosque, Quito. Disponible de: http://sociobosque.ambiente.gob.ec/node/595 [accessed 30 October 2018].

MAE Ministerio del Ambiente del Ecuador (2013) *Proyecto Socio-Bosque*. Quito, Ecuador. Disponible de: http://www.ambiente.gob.ec/programa-socio-bosque

MAE Ministerio del Ambiente del Ecuador (2017) *Tercera Comunicación Nacional del Ecuador sobre Cambio Climático*. Quito, Ecuador.

Martínez, F. (2009) *Influencia de la Textura en la permeabilidad del suelo en la Subcuenca Zamora Huayco – Cantón Loja*. Tesis de grado, Universidad Técnica Particular de Loja, Ecuador.

Mejía-Veintimilla, D., Ochoa-Cueva, P., Samaniego-Rojas, N., Félix, R., Arteaga, J., Crespo, P., Oñate-Valdivieso, F. & Fries, A. (2019) River discharge simulation in the high Andes of southern Ecuador using high-resolution radar observations and meteorological station data. *Remote Sensing*, 11(23), 2804. https://doi.org/10.3390/rs11232804

Ministerio de Fomento de España (2016) *Orden FOM/298/2016, de 15 de febrero, por la que se aprueba la norma 5.2 – IC drenaje superficial de la Instrucción de Carreteras*. BOE núm. 60 de 10 de Marzo de 2016.

Mitasova, H., Hofierka, J., Zlocha, M. & Iverson, L.R. (1996) Modelling topographic potential for erosion and deposition using GIS. *International Journal of Geographical Information Systems*, 10, 629–641.

Molina, A., Govers, G., Vanacker, V., Poesen, J., Zeelmaekers, E. & Cisneros, F. (2007) Runoff generation in a degraded Andean ecosystem: Interaction of vegetation cover and land use. *Catena*, 71, 357–370.

Morales, C. & Parada. S. (2005) *Pobreza, desertificación y degradación de los recursos naturales*. Libros de la CEPAL, N° 87 (LC/G.2277-P/E). Comisión Económica para América Latina y el Caribe (CEPAL)/Sociedad Alemana de Cooperación Técnica (GTZ), Santiago de Chile. 274 p.

NIC Naturaleza y Cultura Internacional (2009) *Protegiendo las fuentes de agua en el sur del Ecuador*. Loja. Disponible de: http://www.naturalezaycultura.org/spanish/htm/ecuador/areas-watersheds.htm

NIC Naturaleza y Cultura Internacional (2013) *Reserva Natural La Ceiba*. Disponible de: www.naturalezaycultura.org/spanish/htm/ecuador/areas-dryforest-laceiba.htm

Nearing, M.A. (1997) A single, continuous function for slope steepness influence on soil loss. *Soil Science Society of America Journal*, 61, 917–919.

Ochoa, P.A., Chamba, Y.M., Arteaga, J.G. & Capa, E.D. (2017) Estimation of suitable areas for coffee growth using a GIS approach and multicriteria evaluation in regions with scarce data. *Applied Engineering in Agriculture*, 33(6), 841–848. https://doi.org/10.13031/aea.12354

Ochoa-Cueva, P., Fries, A., Montesinos, P., Rodríguez-Díaz, J.A. & Boll, J. (2015) Spatial estimation of soil erosion risk by land-cover change in the Andes of southern Ecuador. *Land Degradation and Development*, 26, 565–573. Doi:10.1002/ldr.2219

Oñate-Valdivieso, F., Fries, A., Mendoza, K., Gonzales-Jaramillo, V., Pucha Cofrep, F., Rollenbeck, R. & Bendix, J. (2018) Temporal and spatial analysis of precipitation patterns in an Andean region of southern Ecuador using LAWR weather radar. *Meteorol Atmos Phys*, 30(4), 473–484. https://doi.org/10.1007/s00703-017-0535-8

PAE: Plan Ambiental Ecuatoriano (1995) *Propuesta de políticas y estrategias ambientales. Tercera Propuesta de Discusión*. Comisión Asesora Ambiental de la Presidencia de la República del Ecuador. D: Ing. Luis Carrera de La Torre. CREARIMAGEN, Ecuador.

PEAR (2005) *Plan Estratégico Ambiental Regional Loja-Zamora Chinchipe-El Oro*. Edición Consejo Ambiental Regional CAR, Loja, Ecuador. p. 106.

Pérez-Rodríguez, R., Marques, M.J. & Bienes, R. (2007) Spatial variability of the soil erodibility parameters and their relation with the soil map at subgroup level. *Science of the Total Environment*, 378, 166–173.

Peterson, T.C., Folland, C., Gruza, G., Hogg, W., Mokssit, A. & Plummer, N. (2001) *Report on the Activities of the Working Group on Climate Change Detection and Related Rapporteurs 1998–2001*. WMO, Rep. WCDMP-47, WMO-TD 1071, Geneve, Switzerland. 143pp. Available at: http://etccdi.pacificclimate.org/docs/wgccd.2001.pdf [accessed 30 October 2018].

Programa Socio Bosque (2008) MAE SOCIO BOSQUE. Ministerio del Ambiente del Ecuador. Obtenido de Oojetivos, PSB. Disponible de: http://sociobosque.ambiente.gob.ec/node/173

RAFE Red Agroforestal Ecuatoriana (1998) *Diagnóstico socio-ambiental e institucional de los cinco cantones suroccidentales de Loj*a. *RAFE*. Proyecto Bosque Seco, PBS, Loja. Ec: 159.

Rai, P.K., Chandel, R.S., Mishra, V.N. & Singh, P. (2018) Hydrological inferences through morphometric analysis of lower Kosi river basin of India for water resource management based on remote sensing data. *Applied Water Science*, 8(1), 15.

Renard, K.G. & Freimund, J.R. (1994) Using monthly precipitation data to estimate the R-factor in the RUSLE. *Journal of Hydrology*, 157, 287–306.

Renard, K.G., Foster, G.R., Weesies, G.A., McCool, D.K. & Yoder, D.C. (1997) Predicting soil erosion by water: A guide to conservation planning with the Revised Universal Soil Loss Equation. *Agricultural Research Service (USDA-ARS) Handbook No. 703*. US Department of Agriculture, Washington, DC.

Richter, M. & Moreira-Muñoz, A. (2005) Heterogeneidad climática y diversidad vegetacional en el sur de Ecuador: un método de fitoindicación (in Spanish). *Revista Peruana de Biología*, 12, 217–238.

Richter, R. (2007) *Atmospheric and Topographic Correction for Satellite Imagery* (Atcor-2/3 User Guide, Version 6.3, January). DLR-German Aerospace Center and Remote Sensing Data Center, Germany. 134p.

Rollenbeck, R. & Bendix, J. (2011) Rainfall distribution in the Andes of southern Ecuador derived from blending weather radar data and meteorological field observations. *Atmospheric Research*, 99, 277–289.

Satellite Applications for Geoscience Education (2018) *Oceanography: Ocean Currents*. Available at: https://cimss.ssec.wisc.edu/sage/oceanography/lesson3/concepts.html [accessed 30 October 2018].

Strahler, A.H. (2013) *Introducing Physical Geography* (6th ed.). New York, NY: Wiley. ISBN: 978-1-118-39620-9

Tapia-Armijos, M.F., Homeier, J., Espinosa, C.I., Leuschner, C. & de la Cruz, M. (2015) Deforestation and forest fragmentation in South Ecuador since the 1970s: Losing a hotspot of biodiversity. *PLoS One*, 10. Doi:10.1371/journal.pone.0133701

Témez, J.R. (1978) *Cálculo hidrometeorológico de caudales máximos en pequeñas cuencas naturales*. MOPU, Dirección General de Carreteras. 113p.

Témez, J.R. (1991) Extended and Improved Rational Method: Version of the Highways Administration of Spain. *Proceedings of XXIV Congress IAHR*, Volume A. Madrid, Spain. pp. 33–40.

Thornthwaite, C.W. (1948) An approach toward a rational classification of climate. *Geographical Review*, 38(1), 55–94.

UMAPAL (2005) *Data electrónica de los valores de consumes y las tarifas menduales por categoría comercial del servicio de agua en Loja, período 2004–2005*. Unidad Municipal de Agua Potable y Alcantarillado de Loja, Ecuador.

UMAPAL (2011) *Data electrónica de los valores de consumes y las tarifas menduales por categoría comercial del servicio de agua en Loja, período 2010–2011*. Unidad Municipal de Agua Potable y Alcantarillado de Loja, Ecuador.

USSCS (1972) *National Engineering Handbook* (Sec 4, sup. A). Hydrology, Soil Conservation Service, Washington, DC.

Vuille, M., Bradley, R.S. & Keimig, F. (2000) Climate variability in the Andes of Ecuador and its relation to tropical Pacific and Atlantic sea surface temperature anomalies. *Journal of Climate*, 13, 2520–2535. Doi:10.1175/1520-0442(2000)013< 2520:CVITAO>2.0.CO;2

Windhorst, D., Waltz, T., Timbe, E., Frede, H.G. & Breuer, L. (2013) Impact of elevation and weather patterns on the isotopic composition of precipitation in a tropical montane rainforest. *Hydrology and Earth System Sciences*, 17, 409–419.

Zapata, S.D., Benavides-Muñoz, H.M., Carpio, C.E. & Willis, D.B. (2012) The economic value of basin protection to improve the quality and reliability of potable water supply: The case of Loja, Ecuador. *Water Policy*, 14, 1–13.

Zhang, X. & Yang, F. (2004) *RClimDex (1.0): User Manual.* Climate Research Branch Environment Canada, Ontario. Available at: https://studylib.net/doc/7659063/rclimdex-1-climate-change-indices [accessed 30 October 2018].

Zhang, X., Hegerl, G., Zwiers, F.W. & Kenyon, J. (2005) Avoiding inhomogeneity in percentile-based indices of temperature extremes. *Journal of Climate*, 18, 1641–1651. Available from Homepage: http://etccdi.pacificclimate.org/docs/Zhangetal05JumpPaper.pdf [accessed 30 October 2018].

Annexes

Figures of general illustrations

Map of Ecuador. Loja, study province

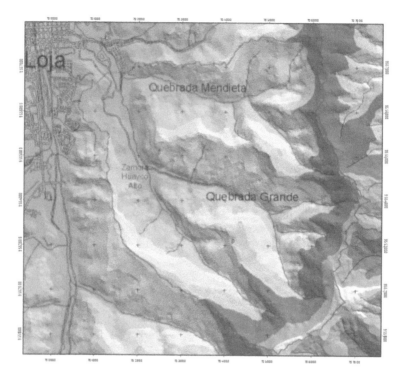

Sub-basin of the Zamora Huayco river

From north to south: Quebradas: Mendieta, El Carmen (Q. Grande) and San Simón.

Source: Dr. Fernando Oñate Valdivieso

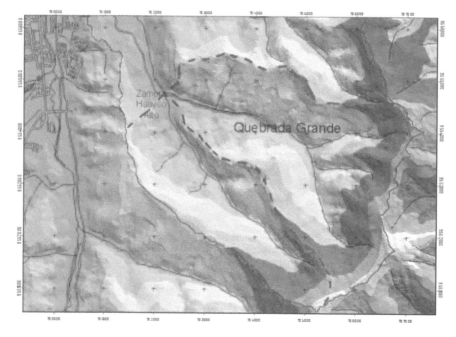

El Carmen micro-watershed (Quebrada Grande) and San Simón

Source: Dr. Fernando Oñate Valdivieso

Photographic annex of El Carmen micro-basin

Medium micro-watershed area destined to the raising of cattle

Channel of El Carmen creek, upstream of the catchment

Medium micro-watershed area destined to the raising of cattle

Part of the forest destroyed in slopes of the micro-basin

Lagoon located on the dividing line of the El Carmen micro-watershed

View of El Carmen stream Geomorphology type "V"

Photographic annex of the San Simón micro-basin

View of the existing infrastructure for the capture of water in the El Carmen creek

View of the sedimentation of the catchment of water in the El Carmen creek

Frontal view of the San Simón catchment weir (on the left, you can see the sand catchers)

Area destined for livestock, waters above the San Simón catchment

Organic cattle waste along a tributary stream

Panoramic view of the micro-basin and a stretch of the San Simón quebrada

Survey model for the drinking water market study

Dear Mr. (a) user, we are carrying out a survey whose purpose is to collaborate with a research to the Lojana community and that as a complement will help us in the realization of our thesis work. All the personal information that you kindly give us will be kept confidential and secret; the purposes are clearly academic.

Poll #: _____ Date: _____

Pollster: _____

Respondent data

Neighborhood streets: _____.
Age: ___ Occupation: Civil Engineering Sex: M ( ) W ( )
Education Level: Primary ( ) High school ( ) Academic (X) Postgraduate ( )
# of people living in your home: _____

Water component

1 Are you connected to the city's drinking water system? Yes ( ) No ( )
2 Yes, previous NO → How do you get water at home? (end of survey)
3 How much do you pay per month for drinking water? US$/month
4 This value looks: Cheap ( ) Normal ( ) Expensive ( )
5 How many hours a day do you have drinking water?
6 Your water meter is: Good ( ) Damaged ( ) Does not have ( )

Open questions

6 What do you consider to be the biggest problem for the drinking water service in Loja?
7 What do you think should be done to solve this problem?
8 What do you think about water rationing?
9 What do you think about the amount of water in recent years?

Perception of water service

10 Today you consider that the amount of water is:

Little ( ) Enough ( ) Much ( )

11 The quality of the water is:

Bad ( ) Fair ( ) Good ( )

12 The regularity of the drinking water service is:

Bad ( ) Fair ( ) Good ( )

13 Do you think that it is necessary to reforest the site of the catchment to maintain a larger amount of water? Yes ( ) No ( )

Why?

14   More than half of the lands belonging to the water supply basin of Zamora Huayco, from where the water is collected to make it drinkable, have private owners, farmers and country men who dedicate themselves to agricultural activities and cause deforestation for converting them into livestock areas. This directly influences the decrease in the quantity and quality of the water. What is your opinion about it?

Monthly income

15   What is your monthly economic family income?
16   How much do you spend monthly for you and your family?

Availability of payment

The streams of San Simón and El Carmen have served to supply water to more than 50% of the Lojana citizens since the 1970s, but due to the population growth in the city, deforestation and degradation of the tributary basin by agricultural and livestock practices, it is estimated that in the future the water will not be sufficient for the consumption of the city and also, the water is exposed to sources of contamination. For these reasons, this market study is developed to support the municipality's task, so that as soon as possible a search plan for the sustainability of the source is established, through reforestation, maintenance and permanent cleaning of the origin of supply, providing protection and improvement of the quantity and quality of the water captured.

The lands of the area where the streams are born (tributary basin) are private property, which are dedicated to agricultural activities; we believe that they should be bought and declared as a protected area of public interest, to intervene and work freely in it. However, there is a lack of sufficient financing to purchase the land, reforest, protect and give adequate maintenance to the water system.

17   With these considerations: How much more would you be willing to pay in the monthly bill of the water service, so that you can buy the lands of the basin and improve the quality and quantity of water through reforestation, protect and conserve water catchments (El Carmen and San Simón basins in Zamora Huayco Alto)?

We appreciate your collaboration and wish you an excellent rest of the day.